THE STORM C
THAN

Since this book was first published, a series of dramatic events have occurred that suggest a sudden climate change may already have started:

- Scientists have found evidence that key ocean currents may be weakening and changing.

- In France, a record-breaking, lethal heat wave claimed thousands of lives.

- Eight-pound blocks of ice fell from the sky in Spain. Scientists found that they had formed in the upper atmosphere for unknown reasons.

- An iceberg twice the size of Delaware broke off the continent of Antarctica. It's the largest ever recorded.

- The National Oceanic and Atmospheric Administration confirmed that the oceans are heating up with unprecedented speed.

- British scientists have confirmed the superstorm scenario.

What will happen next? Has humanity's greatest challenge already begun?

THE COMING GLOBAL SUPERSTORM

ALSO BY ART BELL

The Art of Talk
The Quickening
The Source

ALSO BY WHITLEY STRIEBER

Nonfiction	*Fiction*
Confirmation	The Forbidden Zone
The Secret School	Unholy Fire
Breakthrough	Billy
The Communion Letters	Majestic
Transformation	Catmagic
Communion	The Wild
	Nature's End
	Warday
	Wolf of Shadows
	The Night Church
	Black Magic
	The Hunger
	The Wolfen

Short Stories (Private publication)
Evenings with Demons: Stories from Thirty Years

ART BELL
WHITLEY STRIEBER

THE COMING
GLOBAL
SUPERSTORM

POCKET STAR BOOKS
New York London Toronto Sydney

A Pocket Star Book published by
POCKET BOOKS, a division of Simon & Schuster, Inc.
1230 Avenue of the Americas, New York, NY 10020

ISBN: 0-7434-7065-6

This Pocket Books paperback printing May 2004

10   9   8   7   6   5   4   3   2

POCKET STAR BOOKS and colophon are registered trademarks
of Simon & Schuster, Inc.

Cover design and illustration by Dave Stevenson

Manufactured in the United States of America

For information regarding special discounts for bulk purchases,
please contact Simon & Schuster Special Sales at 1-800-456-6798
or business@simonandschuster.com.

To my fellow humans who will live to see
man's handiwork unfold as nature returns
balance to the planet
—Art Bell

May the children of tomorrow look back on
our era as the one where the healing of
the earth began.
—Whitley Strieber

CONTENTS

CONTENTS

PREFACE

by Whitley Strieber

When Art Bell and I completed this book in August of 1999, we did not expect to see the kinds of enormous changes in climate that it predicts for at least fifteen or twenty years. Nothing in the scientific data suggested otherwise. If anything, climatology was saying that really significant change was decades off.

In fact, we were roundly criticized for even suggesting that the climate might be at the brink of a catastrophe. Matt Lauer of the *Today* show interviewed us, taking a position that was reflected elsewhere in the media: we were irresponsible alarmists attempting to make money by exploiting people's fears. Most national media wouldn't even give us airtime. Not even the cable talk shows would take us.

These refusals came despite the fact that the climate had demonstrated, just a few months before our book was published, that our theories were not only valid, but that the situation was far more urgent than even we had realized.

Our concern was and is that rapid melting of ice at the poles will alter ocean currents and cause a sudden change in climate. This change will start with climactic upheavals unlike any seen before, and will lead to a planetary climate so unlike what we have at present that the long-term disruptions will be devastating.

Sadly, the media rejection of our book meant that the simple, powerful methods of reducing greenhouse gas emissions and thus slowing this process down have not been generally embraced.

People don't want to think about the environmental catastrophe that is staring us in the face because they feel helpless, and they are served by a media that reflects this by being indifferent, or even hostile, to discussing these issues. Worse, the matter has been politicized in the United States, with conservatives following the lead of George W. Bush, Rush Limbaugh, and the Rev. Jerry Falwell, who has gone so far as to state flatly that he "doesn't believe in" global warming. While the Democrats have been more forthcoming, they also perceive public indifference to the issue, with the result that little was made of the growing environmental catastrophe in the 2000 election campaign.

What is so ironic and so sad is that people can, by following the simple procedures outlined in this book, gain very real control over greenhouse gas emissions, and do this without any cost to industry at all. But we need an international effort. Many, many people have to embrace the program for it to work. This was where the media was to have come in. But it did not.

If we do not do something, and right now, sudden climate change is inevitable, and it seems clear that is going to happen much sooner than anybody thought possible. Science is going to be blindsided. In fact, this is happening right now.

On July 11, 2000, Discovery.com published the results of a study from the Norwegian scientific journal *Cicerone*. This story suggested that the northern ice cap would be gone within fifty years. It also pointed to another important factor discussed in our book: Human-generated greenhouse gases are not the only factor causing global warming. They are merely the part of it that we can control. We need to realize that the whole process of sudden climate change is part of the earth's fundamental ecology. As we detail in the book, science tells us that it has happened before, long before there was any "human effect" on the atmosphere. Discovery.com said, "The latest estimates for the ice sheet's end show significantly faster melting than the greenhouse effect can account for."

Just a few weeks later, on August 25, 2000, an astonishing story emerged from a midsummer tourist excursion to the North Pole: The pole had melted.

Where there had been ice from time immemorial, there was now open water. Some scientists were predictably dismissive. On August 25, Mark Serreze of the University of Colorado said on CNN, "We've seen features like this before." He did not say when, however, and in fact, open water has never been observed at the North Pole before, never in human history. The last time the pole lacked an ice cover altogether was the Eocene Era, which ended fifty million years ago.

The key point in this book is that rapid melt of polar ice will cause crucial ocean currents to fail, leading to a sudden and devastating acceleration in climate change.

The ice at both poles is melting at a rate that was completely unexpected even as the book was written in 1999. Worldwide, melting glaciers are unleashing floods of water into oceans that need a certain level of salinity to maintain the stable water temperatures that currents depend on.

Virtually all the world's glaciers, from the Himalayas to the Antarctic, are in retreat. The most serious situation is in Greenland. The Greenland ice sheet is flooding the northern ocean with fifty billion tons of water a year as the ice melts. Worse, the margins of the ice sheet are thinning at a rate of two meters a year. This means that the glaciers themselves are becoming unstable, and the threat exists that they will suddenly slide into the sea.

"We are seeing widespread indications that something like that is going on, causing the glaciers to move faster toward the margins," commented William B. Krabill, author of a study of the glaciers that appeared in the journal *Science* in July 2000. The melting of Greenland would raise ocean levels 21 feet, but major effects on climate will take place long before that happens.

The situation in the Antarctic is just as dire. For the past ten years, the Antarctic ice pack has been breaking up. In March 2000, two massive icebergs broke off the Ross Ice Shelf. The larger of the two was 185 miles long by 23 miles wide, slightly smaller than the state of Connecticut. In September of 2000, Professor John Lowe of London University suggested that global warming could cause a mini–ice age in Britain within a few decades. He described the evidence as "startling." This is not an isolated event. It is one of a series of huge ice breaks in the Antarctic that began in 1987, when a berg the size of Rhode Island broke off the Ross Ice Shelf. On February 9, 1988, the *New York Times* reported "an extraordinary two years of glacial breakup." This led to what is now the complete destruction of the Larsen Ice Shelf, and the increasing instability of the Ross.

The melt of floating ice, such as the polar cap and the Antarctic ice shelves, will not cause sea levels to rise, and there is not much evidence that a sudden glacial surge off Greenland or the Antarctic will take place in the near future. The issue is not sea levels, it's fresh water: Polar melt is flooding the northern and southern oceans with it.

This is exactly what is believed to have led to

the last great climatic event, which we discuss in detail in this book. We also discuss the substantial evidence that this event began with a ferocious series of storms that led to two hundred years of reduced temperatures in the northern hemisphere, and a reduction of temperatures above the Arctic Circle that is just now ending.

There appears to be a great climatic cycle that is marked by long periods of stability punctuated by sudden change. We are almost certainly at one of the points of sudden change. It will come when fresh water flooding into the polar oceans combines with rising air temperatures to increase water temperature so much that the great currents that control climate stop flowing.

When wrote our book, we thought that this situation would develop over a period of years. Given the past twelve months, however, it now seems clear that it could happen at any time. Indeed, it has probably already started.

On November 27, 1999, the prestigious British newsweekly *The New Scientist* published a story entitled "Freezing Future: There's Now Alarming Evidence that Europe Is Facing an Ice Age." The story went on to corroborate the warnings voiced in our book. "The ocean currents that give Europe its mild climate are changing," *The New Scientist* said. "Scientists have found evidence that global warming may cause a big freeze by switching off a current called the North Atlantic Drift."

Due to the flooding of the northern oceans by

fresh water from rapidly melting polar ice, there have already been dramatic shifts in the very currents that control our weather. There is even data showing that a deep sea current in the arctic has actually gone into reverse.

When we read these stories, shortly before our book was published, we immediately informed the book's publicist, who sent out press releases.

For the most part, the media did not respond, not even to this open and obvious physical evidence that our warnings were not only appropriate, but frighteningly timely.

Not all media powers reacted the same way. Columnist Liz Smith was particularly affected by the book. "I picked up this book just as those New Year storms were savaging Europe, leaving almost 100 people dead. But those storms are only a drop in the bucket compared to what this book claims is in store for us unless we get busy and do something about it."

The storms she referred to were truly alarming natural events. Over a hundred people were killed as winds reaching as high as 125 miles an hour ravaged Europe in two successive storm waves. The second storm, on December 28, 1999, swept France with winds of unprecedented velocity, destroying hundreds of millions of trees and wreaking havoc with power lines, roadways, and homes.

Prior to this, storms from Spain to Italy reached incredible altitude, resulting in falls of eight-pound hail. Why these blocks of ice were plummeting from the skies mystified scientists. "The most surprised person of all by this phenomenon is me," said geologist Jesus Martinez Frias, who was heading a team of scientists collecting and analyzing the blocks of ice. Months later, they announced that the ice was indeed a weather-generated phenomenon.

To us, it was just a precursor of things to come. It meant that cloud tops were reaching the kind of extreme heights predicted in this book, and allowing ice masses to form at ultrahigh altitudes, then grow to immense size as they plummeted to the earth. This is happening because temperatures near the surface are rising, due to heat being trapped by greenhouse gases. Temperatures above the stratosphere are dropping, due to the fact that less and less heat is radiating up from below.

The effect was reported by scientists in May 1999. The mesosphere, about forty miles up, has been cooling by a degree a year for the past ten years. This is ten times faster than anyone had previously predicted. Gary Thomas of the Laboratory for Atmospheric and Space Physics at the University of Colorado says that this may be the climate change "miner's canary." In the past, miners would take a canary with them into deep mines. The bird would die from odorless gases long before the miners themselves would realize it was there. If the canary died, the miners got out.

The radical temperature difference between the upper and lower atmosphere means that the potential for extreme storms is rising, and throughout 1999 and 2000, they marched across the world. Not only were weather extremes felt in Europe, there were devastating storms in Venezuela, China, India, and Southeast Asia. When torrential rains flooded Caracas on December 20, 1999, 30,000 people were killed. The previous May, an Oklahoma tornado set a world wind record when wind speeds in the funnel cloud were recorded at 318 miles per hour. In November 1999, a monstrous supercyclone very much like the one described in this book killed at least 12,000 people in India. Millions were left homeless, and desperate survivors began burning the carcasses of 200,000 drowned cattle, buffaloes, and pigs in an effort to prevent the spread of disease.

On November 23, 1999, the *New York Times* published an article that directly supported the theories in *The Coming Global Superstorm*. The article described the same climatic event that is discussed in this book, the one that brought on weather chaos toward the end of the last ice age. Dr. Gerald R. Dickens of James Cook University in Australia compared the way a warming trend like the one we are in now climaxes to a rubber band: "You gradually pull at both ends and, at some instant, the rubber band suddenly breaks."

The article, for the most part, discussed a sudden climate change event that took place 55 million years ago. But it illustrated our point very clearly. In the words of Dr. Dickens: "The earth can, for natural reasons, suddenly change dramatically."

When you add the push effect of human intervention to the natural processes that are occurring right now, there is obviously potential for explosive change.

It is now October 2000, and we have seen a devastating cascade of changes, most of which have been totally unexpected. The weather has become ultraviolent, highly unpredictable, and extremely dangerous.

On January 31, 2000, *U.S. News & World Report* published an article about weather change, essentially blaming the present situation on short-lived oceanic phenomena like El Niño and La Niña, alternating warm and cold water cycles that occur in the Pacific Ocean and profoundly affect climate in the Americas. The article concluded, "It's too soon to panic."

Too soon? We don't think so. As this preface is being written, world weather patterns are sliding deeper into chaos. Over five million acres of arid land have already burned in the American west, and predictions are that six million acres will have burned by the end of the year. On August 27, 2000, the Montana fires combined into a single, giant 247,000-acre conflagration and became the largest single fire ever recorded in the United States. Then the winds began to rise.

At the same time that fires raged in the west, the midwest and northeast were experiencing one of their wettest summers on record. This pattern of extremes was a continuation of a situation that began to evolve in January, when temperatures plummeted to -15° in New York, while they soared to 81° in south Texas.

Regional extremes like these continued worldwide, and the summer of 2000 promised, like the summers of 1997 through 1999, to be one of the hottest on record. In early July, a thermal invasion of extremely hot air from the Sahara Desert invaded southeastern Europe and Turkey. Temperatures exploded to unprecedented heights, shattering records across the region. Turkey and Greece saw temperatures as high as 113°, and it went to 111° in the Balkans. Roads became impassable because of melting tar, and fires broke out across the region.

In India in September of 2000, fifteen million people were made homeless by monsoon flooding.

A more disturbing event took place in northern England on August 21. Freak storms, not predicted by British meteorologists, surprised Hull and York with torrential rain, hail, a serious tornado, and five inches of snow. Similar weather struck in North Wales two hundred miles away, where hail blocked roads as temperatures plummeted.

This is exactly the sort of bizarre weather that would take place in this part of the world if the normal flow of warm water from the south was slowing down, but had not yet ended altogether.

Our book predicts these effects, but it goes farther than that: it predicts what is in store for us over the next few years, and it would appear that climate change during this period could be far more dramatic than expected.

The book was not only shunned by the media in the United States, it received review attention that was almost entirely hostile. This was in dramatic contrast to the British press, which generally supported the book.

However, no major reviews anywhere in the world discussed our speculations that there was almost certainly a much more developed society on earth far earlier than conventional scientific wisdom.

Throughout the year 2000, there have been a series of discoveries pushing the dates for crucial human developments back farther and farther, much more in line with our theories.

We suggested that this earlier society was consumed by a previous dramatic weather upheaval. We made our case that this society must have existed by briefly surveying a number of ancient structures whose origins and structural engineering have not been satisfactorily explained. Among these structures was one off the small Japanese island of Yonaguni in the Pacific. This enormous structure subsided into the ocean about nine thousand years ago, approximately during the time that the last upheaval was taking place.

Despite compelling evidence that it is man-made, science has dismissed it as of geologic origin, largely because of its dates. There could have been no civilization in Japan nine thousand years ago that would have been capable of building such a large structure.

In November 1999, however, unmistakable carvings of human origin were found on the sea floor near the monument. It has now been theorized that the object must have come up out of the sea about five thousand years ago, then subsided again after the carvings were made.

This seems unlikely to be the last word on this particular structure. Nevertheless, our speculations about the past were just controversial enough to justify our entire thesis's being ignored. This means that readers did not get informed either of the crisis nature of the present situation or of the means we have of exerting what could be a crucial degree of human control over it.

Whether the civilization we believe may have been there existed or not, conventional science certainly offers ample and all but irrefutable evidence that the climatic upheaval that took place then was caused by a "snap back" of the climate from a global warming spike *exactly* like the one that is taking place now.

There is substantial evidence, also, that the upheaval overtook this planet during a single, awful season of incredible weather chaos. If such an event were to happen today, it would result in the deaths, ultimately, of many billions of people.

Not only would the initial catastrophe cause many to die, the agricultural disaster that followed would lead to a much greater loss of life.

On July 11, 2000 in the *New York Times,* Bob Herbert's "In America" column quoted Dr. Michael Oppenheimer, chief scientist of Environmental Defense, a national environmental research organization. Dr. Oppenheimer said, "The last time it was as warm as it probably will be by the end of this century, globally speaking, was several million years ago. There is no way we can know for sure if that's a world we can safely cope with."

Given the fact that Dr. Oppenheimer did not know, at that time, that the North Pole would melt in a few weeks, it is safe to say that he may not have been talking about the end of this century at all. His remarks might be even more accurate if he had said, "at the end of the next few months or years."

The Coming Global Superstorm is not a book of speculation about some vague possible future. It is a call to action about events that are almost certainly staring us right in the face. It seems now, as of August 2000, that the kind of climate change we describe is much, much closer to happening than we ever dreamed when we were writing the book . . . and that was just one year ago.

—WHITLEY STRIEBER
September 26, 2000

Prologue

The Storm Begins

The earliest warning sign was something so small that it was hardly noticed at all.

The National Data Buoy Center's buoy 44011, anchored off Georges Bank 170 miles east of Hyannis, Massachusetts, appeared to be sending a faulty signal. That was the only sign from any scientific instrument anywhere in the world that two billion human lives had just come into mortal jeopardy.

The warning should have come weeks earlier, could have come years earlier. There were climatologists who were concerned enough to have begun studies that would lead to the deployment of a warning system. But there was no budget. Congress, mired in its false debate about whether global warming was even happening, wouldn't pay for any studies of the flow of the North Atlantic Current, even though it is the lifeblood of our world.

What happened off Georges Bank was this: The water temperature reading from this six-meter Nomad buoy fell suddenly from 48.1 degrees Fahrenheit to 36.3 degrees. This is a huge drop in seawater temperature to happen overnight, and it caused the National Data Buoy Center to list the buoy as malfunctioning. The issue was noted, and a bulletin was distributed within the National Oceanic and Atmospheric Administration to the effect that water temperature readings from this buoy were to be disregarded until after routine maintenance was next performed.

This standard notice never reached anybody who might have been concerned about its true meaning.

A few days later, another buoy appeared to malfunction. This one was part of the Global Ocean Observing System, feeding data to the Australian Oceanographic Data Centre from its station in the Southern Ocean a thousand miles from the Antarctic. Operating under the protocols of the Global Temperature-Salinity Profile Program, AODC transmitted the data to Canada's Marine Environmental Data Service. Again, the failure of a buoy was duly noted, but the maintenance bulletin didn't reach the same people who'd seen the one for the buoy off Georges Bank. Why would it? Maintenance of the Antarctic buoy would be performed by the Australians, not the Americans.

Mankind's greatest civilization now had only a few weeks to live.

Had the scientists working on the Atlantic Climate Change Experiment known what had happened, they would certainly have been alarmed. As it was, their plan to release one hundred subsurface drifting buoys to study the North Atlantic Current was still in the preparation stage, still waiting on funding.

Even though there was no source of data to sound the warning that the world's greatest ocean current had just changed its route, it wasn't long before people from Sydney to Tokyo, from Vladivostok to Dusseldorf, from London to Los Angeles, knew that something had gone terribly wrong with the weather.

New York had been experiencing the warmest February on record. Temperatures were reaching their highest levels ever recorded for the month—91 degrees Fahrenheit.

Once, people would have been laughing. Nobody was laughing now.

Across the whole southern coast of the United States, from Brownsville, Texas, to Cape Fear, North Carolina, an unusual southerly flow of air began. Tender young leaves shuddered on early sprouting trees in south Texas. In Mississippi, ancient oaks tossed and bowed. Along the Carolina coast, the wind hissed through pine forests. In the warm, winter-naked northeast, clattering

*limbs and moaning eaves made it sound
cold. But it was not cold. In fact, temperatures and humidity were rising. As far as the
United States was concerned, even though it
was the dead of winter, summer had begun.*

*In Australia and New Zealand, the oppo-
site happened. The austral summer, which
had been fairly normal through January,
began to show signs of an unexpected
change in February, when snow now began
falling in the mountains of New Zealand's
southern island. Record cold gripped
Auckland. Australia, farther north, remained
locked in record heat, but it was clear that
this would soon change.*

*At the Russian Federation's Meteorological
Data Processing Center at Obninsk, an
image was picked up off a high-density data
stream from an orbiting ENVISAT satellite
that confirmed what ground observers were
reporting: an extremely unusual storm had
suddenly formed over the Russian Arctic.
Weather systems like this had been seen only
a few times before. The first one, which had
formed over Duplin County, North Carolina,*

on the night of April 15, 1999, had been dubbed the "tornadocane." It was a massive tornado-producing supercell with the circulation characteristics of a hurricane. Winds in the system had reached 165 miles an hour. It had even formed an eye in an area near the mesocyclone, or tornado-producing region of the storm.

Instantly recognizing how unusual the storm they were seeing was, the Russian scientists reported it to the World Meteorological Organization. China's FY-1 Polar Orbiting Meteorological Satellite Program was also watching the storm's development. They sent the WMO an urgent message: The storm's CAPE, or collective available potential energy, appeared to be rising at a very high rate.

What a storm like that was doing there at this time of year, nobody knew—let alone why it was becoming so powerful.

All across southern Europe, from Madrid to Istanbul, a hard, dry wind began roaring up from the south. In New York, low, wet clouds had been swarming northward for two days. In Atlanta, average wind speeds had reached thirty miles an hour. In Houston, the average speed was forty.

All over the world, meteorologists were watching the situation. So far, however, nobody had connected what was happening in different parts of the planet. Thinking was still highly localized, although numerous research facilities were observing the data being transmitted by the Russian and Chinese satellites.

Then a typhoon appeared in the central Pacific. It formed over a matter of hours—faster, in fact, than had any typhoon ever previously recorded. Inside of a week, this massive storm was menacing coastlines from the Philippines to Japan. It was graded a Category 4 storm on the Saffir-Simpson scale, and declared a supertyphoon. It was called Max.

The U.S. National Severe Storms Laboratory, recognizing the extraordinary power of this storm, began to acquire data on it from all available sources. Close to the center of the system, wind gusts were exceeding two hundred miles an hour. Emergency weather bulletins went out across the whole of the Pacific.

Meanwhile, the Australian Bureau of Meteorology was observing another kind of system on the high seas south and west of Tasmania. This system was moving on a track that had never been seen before.

They also reported this storm to the World Meteorological Organization. Realizing that it was now receiving data on three extremely unusual storms in different parts of the world, the WMO appealed to the U.S. National Severe Storms Laboratory for help in interpreting the situation.

With wind speeds now reaching 200 miles an hour, Max was raised to Category 5. There was a possibility that it would become the strongest storm ever recorded. The "tornadocane" over the Russian Arctic was becoming part of a system of similar storms that appeared to be forming with the North Pole as their rough center.

But in Paris, temperatures were rising toward the nineties. In New York and

Toronto, southerly winds in excess of forty miles an hour were being recorded.

A supertanker, the Exxon Invincible, reported that it was taking on water off Cape Race, Newfoundland, and in danger of breaking up. From Newfoundland to North Carolina the alert was sounded: The area was in peril of the greatest oil spill in history.

In Dallas, you could smell the salt tang of the Gulf of Mexico three hundred miles to the south. In London, temperatures, which had been reaching through records, had finally begun dropping. Across Europe, storms began to crash and roar, and the nights of fifty cities were streaked by lightning.

By now, climatologists and meteorologists worldwide were aware that the planet's weather was in upheaval. At the U.S. National Severe Storms Laboratory the crucial question was first asked: Why?

1

Present Danger

~~~~~~~~~~~~~~~~~~~~~~~~~~~~~~~~~

NINETEEN NINETY-NINE WAS THE MOST VIOLENT
year in the modern history of weather. So was
1998. So was 1997. And 1996. Anybody who
glances at a weather report from time to time can
see that something extraordinary is happening.
But exactly what that is remains a matter of controversy.

For twenty years, we have been bombarded with
warnings that global warming is a real and present
danger. Equally, there have been claims that it's all
nonsense.

On March 15, 1999, scientists at the University
of Arizona and the University of Massachusetts reported on their construction of a thousand-year
record of earth's average temperature. The results

were shocking. What has happened is that a nine-hundred-year-long cooling trend has been suddenly and decisively reversed in the past fifty years. Due to the rise in heat-trapping greenhouse gases, ferocious warming is under way. The scientists predicted that the earth will shortly be warmer than it has been in millions of years.

A climatological nightmare is upon us. It is almost certainly the most dangerous thing that has ever happened in our history. However, there is a surprising amount that we can do about it. Some of it involves personal action. Some of it involves the whole society. None of it is particularly difficult or expensive, and none of it will place a cost burden on government, business, or the individual.

How effective will it be? That remains to be seen. So far, the fact that we cannot answer the question of just how dangerous global warming actually is, has meant that nobody is doing anything very decisive. But the situation is getting more and more serious. It has become clear that the deterioration of the atmosphere—indeed, of the whole biosphere—is happening a lot faster than even the most concerned climatologists imagined just a short time ago.

What does this mean? What might happen? We must find a way to understand. We must, because we have to empower ourselves to prevent it. Could it be that the worst climate disaster of all—an event barely whispered about—is actually happening right now? Could we be at the edge of runaway climate change—an event so devastating that it could abruptly leave the world unable to feed itself, perhaps even visit it with unimaginable destruction?

To find out, we must take a journey not only through the shocking record of current climate change, but also into the amazing history of the world's weather.

At this point, almost any violent change in climate will batter our civilization because it is so enormous and makes such a massive demand on the environment. Even the unthinkable could happen: our civilization could fall.

Earth's climate works like a rubber band being stretched and suddenly released. For years, eons even, the stresses slowly build as the chemistry of the air changes. And then, in a matter of a few years or even a few months, there is a shift so vast that we can scarcely begin to imagine it.

Earth, it seems, has a powerful regulatory mechanism built into its climate. Heat increases to a certain point, and then the whole system breaks down. Cold air comes roaring down from the north, flooding the previously overheated Northern Hemisphere.

Suddenly, a new era of cold weather begins. We know, generally, how this happens. But not even science has as yet faced the fact that this change must be accompanied by an absolutely massive release of energy, as earth's climate strives to reorganize itself. In other words, this great shift of climate is almost certainly accompanied by a great storm or series of storms, a weather upheaval outside of contemporary human experience. We believe that it has happened before, and that traces of what we are calling the superstorm exist in the fossil record. We believe that it comes on suddenly and that it is so destructive that it has the potential to end our civilization.

These are sensational claims, but we can prove that nature pulls the trigger suddenly and, therefore, that the rebalancing of the climate that follows must also be very sudden and involve titanic energies. This suggests that our present situation may be extremely perilous.

Over the past three million years the earth has been locked in an unusually harsh climate system. During this period, our climate has flipped from warm to cold conditions and back again many times. Again and again, earth has warmed up, getting hotter and hotter until—very suddenly—the glaciers have come back and entombed a quarter of the planet in ice for upwards of a hundred thousand years. Sometimes, the cooling event has not resulted in a long-term buildup of ice. Sometimes, as happened around 8,000 B.C., sudden cooling has not led to the return of the ice, but has only interrupted the warming process for a short time.

All of the factors that have caused sudden climate change in the past are lining up right now. This change, which we will show is part of a vast natural cycle, has been sped up this time by human activity. When the change comes, it is likely to be much more violent than ever before, and we will offer evidence from recent and unexpected climatological data that indicates why this would be so.

We will look at the last great upheaval through the eyes of the people who were living then. Examining the fossil record, we will identify the season in which it took place. And we will see why that particular event did not result in a new ice age and learn exactly how to tell if the changes the next one brings will cause one or not.

What will this climate change be like for you and your family? This depends on where you live. The farther north your home, the more likely you will have to move quickly south.

When the warm ocean currents that now flow north cease to do so, our whole climate will change. It is our contention that the energy necessary for the superstorm will be created at that time.

Say you live in Dallas or Madrid or Rome. Your first indication that the superstorm is building might be weather reports to the effect that a series of cold fronts are moving down from the Arctic, one after another. This could happen at any time of the year. You would hear that more northern places—Toronto, Stockholm, Beijing—were receiving extremely heavy weather—extraordinary rain in the summer, unprecedented blizzards in the winter. This would continue for a week or more, always building in intensity. Across the northern plains of the world—the American High Plains, the central Asian steppe—wind gusts of upwards of one hundred miles an hour would start to be recorded. We believe that it would get worse, and we will make our case over the course of this book.

Places like Edmonton and Semipalatinsk, then Minneapolis and Moscow, would cease to communicate with the outside world. Alaska and northern Siberia would have gone silent before.

From Europe to Asia to America, whole populations would be desperately attempting to move south. Because the same changes that affected currents in the North Atlantic would alter the movement of currents in the Southern Hemisphere, Australia and New Zealand would also be affected. There, summer would have turned to winter, or normal winter would have become extremely cold. Heavy seas would devastate the southern coasts of the continent. Typhoons, blowing up suddenly, would smash into the Philippines, Japan, and the Pacific islands.

The farther north you were, the more extreme conditions would be. Day after day, the storms would continue, becoming more complex and organized, larger, taking on forms never observed before.

All over the Northern Hemisphere, massive population movements would be taking place. There would be mass disorganization, and many, many people would be overrun by the superstorm.

After the superstorm was over, it would gradually become clear that a catastrophe of breathtaking proportions had occurred. The only reports from Europe would be coming from Portugal, southern Italy, and southern Spain. The entire American Midwest would be under a sheet of ice, one that would extend across Siberia and northern Europe as well. This ice would reflect vast amounts of sunlight and heat back into space.

If the storm—as the last one appears to have done—hit in summer, the ice would probably melt. It is possible that this happened the last time and, as we shall see, was recorded in myth all over the world.

If the storm took place in the fall or winter, then the ice could conceivably compress so much in the next few months and reflect back so much heat and light that the next summer simply would not be warm enough to melt it. The winter that followed would be the coldest in history.

The ultimate and ironic effect of global warming would have become clear to the survivors: a new ice age would have begun.

2

Our Mysterious Past

~~~~~~~~~~~~~~~~~~~~~~~~~~~~~~~~

THE SUPERSTORM HAS ALMOST CERTAINLY HAP-
pened before, probably many times. The last such in-
cident may actually have been recorded by early
man, who may have left us a description of it and a
set of warnings that must not continue to be ignored. However, we have long assumed that early
human legends, when they refer to events we regard
as fantastic, must simply be acts of the imagination.

Maybe that's unwise, and maybe we should take
a more careful look backward, to see what our
ancestors may have been trying to tell us. Learning
from history is not something that our society
does well. We prefer to rely on science and to look
to the future. In this case, though, the past may
have sent us a message that is vital to our survival.

The established story of the human past is this: For the last two million years, prehuman species such as *Homo erectus,* and later Neanderthal man wandered the African landscape, spreading slowly into Europe and Asia. They used crude tools. We know this because we have found them. They did not have very well developed language. We know this because their necks were too short to support the kind of breath control that complex speech requires. They could have uttered only simple words. Then, about a hundred thousand years ago Cro-Magnon man appeared. He was built very differently from his predecessors. He stood tall and had a broad, light skull without heavy simian browridges. With his long neck, he did have the capacity for complex language, which was one of the things that made it possible for him to develop civilization.

About seven thousand years ago, the first towns began to be built, followed by the first cities a thousand or so years later.

As recently as 1995, that was scientific dogma about the past, backed by many years of careful archaeological research.

Recently, however, controversial researchers such as Graham Hancock, Richard Thompson, and William Corliss have begun urging us to take a harder look at our own past. Hancock, in books such as *Fingerprints of the Gods,* has suggested that ancient civilizations must have been far more scientifically capable than we have thought. Corliss, with his *Sourcebook* series, has delved into abandoned and ignored research to unearth hundreds of unexplained discoveries, in the process gently reminding the scientific community that it has a bad habit of dismissing what it can't explain instead of evolving better theories.

Mainstream science, of necessity more conservative than those whose work is at the speculative edge, is finally beginning to respond to the questions they have been asking.

It is at last being realized that ancient man may have been a careful observer of his world and, therefore, that the stories he has bequeathed to us in the form of myth and legend may not reflect just a primitive imagination, but actually offer observations that are vitally important to us now.

But before we get to the human past, we need to go *way* back, all the way to the first few billion years of earth's existence, long before a single living thing existed.

Our reason for discussing this seemingly unrelated subject, if only briefly, relates to something very strange that we think we have found hidden in human history. We believe that we have detected a level of planning in the evolution of culture that is totally unremarked anywhere else. If this represents a message left for our age by our ancestors, to understand it and why someone in the deep past might have had an overwhelmingly powerful motive for trying to help us now, we need to understand something quite unexpected about ourselves, and we must begin by returning to the very origin of the earth itself.

At that time, what was to become the earth was a glowing dust cloud centered on a molten ball about half the size of the current planet. For eons this mass had been orbiting the sun, growing slowly larger as it cooled and more and more of the dust got sucked into the ever darker, more planetlike center.

Then something happened—something that was improbable but not impossible, especially not in the comet-ridden murk of the early solar system. An enormous object crashed into the ball of rock and lava we now call Earth. Within moments, it became a double planet. The smaller body orbited the larger, at first quite quickly. But as the smaller object gradually moved away, its orbital speed also dropped.

The impact that had created it was huge: we now call the crater it left the Pacific Ocean.

So we ended up with a planet with a huge moon that was orbiting it more and more slowly. Over time, the gradually increasing drag of the moon's gravity slowed down the rotational wind of the earth, which would otherwise blow in excess of two hundred miles an hour.

The balance of the earth-moon system is exquisite. Were it not for the moon being just the size that it is and orbiting the earth just as it does, nothing more complex than a lichen would ever have evolved here. The march of life on earth would never even have started.

So now we have two improbable events. First, the moon broke off from the earth without the whole planet being pulverized. Second, the way it ended up orbiting earth slowed down earth's rotational wind, creating a weather environment in which higher life-forms could develop.

One improbable event can be put off to chance. So can two—maybe. But there are more—many of them. One example is the Jupiter Effect. If not for the fact that Jupiter is the size that it is and has a perfectly circular orbit, earth would not be as far from the sun as it is. Even fifty thousand miles closer, and we would be outside of the envelope of livability. Earth would be too hot. A few thousand miles farther out, and the planet would be frozen.

The universe is so enormous that the earth-moon system, as improbable as it is, might still have come about by chance. However, it must be incredibly rare. It is certainly the only one of its kind in our own solar system. As low wind velocities are essential if large surface-dwelling creatures are going to evolve, a planet without a moon like ours could never sustain us. In fact, not even insects could evolve on a planet without a moon orbiting it in such a way that the tendency of the atmosphere to be "dragged" up to a velocity that is a significant percentage of the planet's orbital speed is dampened.

How many such planetary systems might there be in the universe? Probably not very many. Maybe very few indeed. At best, the dusting of intelligent species scattered across the universe are probably profoundly, incredibly alone.

But there are more things that make higher life forms rare. First, most galaxies appear to be what scientists call "gamma ray bursters." This means that they periodically emit hugely powerful explosions of gamma rays that sterilize the whole galaxy with radiation so intense that higher life forms like plants and animals—let alone intelligent creatures—could not survive or even evolve.

So most galaxies are probably dead.

There's even more, though. Our sun is a yellow dwarf. If it was a larger star, it probably wouldn't have a zone around it that was congenial to the evolution of life. There would be too much radiation involved. A smaller star wouldn't emit enough heat.

Yellow dwarves, however, are pretty common, which would be a plus for life if only so many of them did not emit huge solar flares. If our sun was like most of the yellow dwarves that have been studied, planet Earth would be a blackened hunk of rock. Most of them expel devastating flares all the way out to the equivalent of the orbit of Jupiter.

There is also something about the evolution of species that suggests that we may be extremely rare.

Extinction events—seemingly random—have played a huge role in the evolution of life on earth. But there is something about these events that is not understood that is crucial to our understanding of ourselves.

An example of what this is can be seen by examining what happened after the event that killed the dinosaurs. As many as seventy-five percent of all species died, including the vast majority of all animals. For every thousand creatures alive before the event, only ten were alive afterward. And yet the same thing that has happened after all the other events happened again: earth didn't become sterile, and life didn't have to claw its way up from nothing.

Instead, the devastated landscape burst forth with new creatures, and in ten million years it was teeming once again with animals—better animals than before. The new creatures were smarter, stronger, and more adaptable than the ones that had been destroyed.

Every time it takes a hit, it seems that the earth comes back as a better model. The fossil record demonstrates this clearly. It is hard to see how chance can be the only thing at work here. The earth-moon system is a life-creating machine, and the periodic extinctions that take place seem to actually accelerate the process of evolution.

When you add this effect to the lucky chance of the way the moon orbits the earth and the Jupiter Effect, you have to ask if there is enough chance to explain it all. Or if *any* amount of chance can explain the way that mass extinctions seem to lead to better, more efficient creatures replacing the ones that were destroyed.

It is impossible to generate a statistical probability for the evolution of mankind, because of all the unknowns involved. But it must be very, very small.

Carl Sagan, who was an advocate of the idea that there must be billions and billions of intelligent species in the universe, did not consider the need for a moon to slow down rotational winds, or any of the other improbabilities we have been discussing. As a result of society's deep assumption that the universe must be teeming with life, we may have devalued ourselves. We may not realize how rare we actually are, and therefore fail to understand how terribly, terribly important it is that we always err on the side of caution when we are dealing with possible harm to the human species. We must not tempt extinction. We must not experiment with our children's future.

Perhaps somebody in the distant past knew more, and in understanding how rare we actually are, cared more for us than we do for ourselves.

When we face how alone we probably are, we see ourselves in a new way and we begin to understand the breathtaking urgency behind a project like *The Coming Global Superstorm* and the motives that might have led our ancestors, trapped in the maelstrom of just such an event, to try to leave a warning about it for future ages.

We are about to suggest that there may be a civilization, or something very like one, lost deep in the human past, that has tried to transmit a message to our era. We know that our suggestion is speculative, but, as we will show, ideas about the human past are under radical revision at the present. And lest we paint a picture of conventional scientists as being hidebound and unwilling to change, we must note that some of the most exciting revisionist thinking about the human past is coming out of the established scientific community. In fact, discoveries far more incredible than even the most dramatic speculations put forward by people like Hancock are not even in serious scientific dispute. But we are still far from realizing that the past may have understood enough about itself and the inner nature of man to have been capable of sending a warning across at least eight thousand years of time, from an era that we have always assumed was primitive.

It's urgent, however, that we find this message if it is there, because the winds of time are blowing harder every day, and it will not be long before we are going to need shelter from their power. For impacts from other celestial bodies are not the only things that have caused extinctions on earth. Climate plays a huge role, maybe the most important one of all, and the past may have vital information for us about what that means right now.

3

Trouble in the South

"*Today will be fine, with a high of twenty-two in Sydney, cooler at the beaches.*"

That's what they'd said, and that's what the crowds along Bondi Beach expected. At the Australian Meteorological Bureau, however, they were still watching the unusual storm system to the south, but it was hanging well offshore. The seasonal flow of warm air off the continent, they felt, would keep it well off the coast until it dissipated.

It did not happen that way, not this time. By noon, it was clear that the weather forecast was off. Thunderstorms were becoming visible just above the horizon. The surf was getting up along beaches as far south as Wollongong. From places like the top of the AMP Tower, the view to the south revealed roiling, violent clouds. Most of the tourists were unaware of how unusual this was, but the locals—the guards and attendants on the observation platform—regarded the sight as something extremely strange. It was not right. It was ominous.

At the Meteorological Bureau, the weather forecast was being changed. Normal airflow had collapsed in a matter of hours. The pattern was now characteristic of July— the dead of winter for Australia. The AXM Radio announcer was handed a new forecast: ". . . turning colder by evening, with temperatures ranging down to fourteen. Possibility of rain and gusty winds after sunset."

The storm swept into an astonished Sydney, bringing with it a rocketing drop in temperatures. By seven, the temperature was an amazing 12 degrees Celsius at Kingsford-Smith Airport. Freezing rain swept through the streets, emptying them of crowds. Confused tourists, lacking cold-weather gear, huddled in their hotels. Sydney's normally active nightlife came to a stop. Local people, knowing for certain now that something extremely strange was happening, began to go home. Darling Harbor emptied. In fact, the whole center of the city became a ghost town.

Rain swept the streets; wind raced through the girders of the "Coathanger," the famous Sydney Harbor Bridge, and screamed eerily in the complex eaves of the Sydney Opera House, a sound rarely heard but never forgotten. Some joked that this sound was the ghost of the great Australian diva Nellie Melba mourning the loss of her voice.

Restaurants began to empty when the rain turned to sleet. At four minutes of nine, the first fatality of the storm occurred—an Indonesian tourist, inexplicably wandering around on South Head's slippery flanks, got *swept away by the wind and fell into the wild ocean.*

Publicly, the Australian Meteorological Bureau attributed the strange weather to an unusual frontal system that was a result of the powerful typhoon that was now approaching Japan. Behind the scenes, though, the whole meteorological world was in crisis, not only Australia's. The weather in the Northern Hemisphere was far worse, in fact, and continuing to deteriorate. Moscow was reporting snow, as was Beijing. Enormous storm complexes were still developing in the Arctic, centering now over the eastern polar sea. Just weeks ago, the fact that the North Pole had been replaced by blue water for the first time in a quarter of a million years had made headlines.

This seemed far from southern Australia, however, and local problems were beginning to capture everybody's attention. Word from New Zealand was grim: South Island was in the grip of the worst blizzard in its history. Anyone could see that the same huge, inter-linked mass of storms was responsible both for the situation there and what was happening in Victoria and New South Wales.

The strangest thing was that the Antarctic Automatic Weather Station Project, which supported a network of weather stations across the whole of the Antarctic continent, reported nothing unusual. Temperatures were normal. There was no strange storm activity on the continent itself.

But satellite photography told meteorologists in Sydney that an even stronger storm was building and extending out of the one that was causing all the damage in New Zealand. In fact, a new frontal line had developed along a route that had never been recorded before. It was longer and stronger than anything previously seen, a great, roiling mass of rain and snow squalls that extended for two thousand miles across seas that, at this time of year, ought to be warm.

The Antarctic Meteorology Research Center could offer no more of an explanation than was available from the automatic stations and satellites. It was easy to tell what was happening—summer had suddenly turned to winter in the Southern Hemisphere.

But why? Why was a winter storm packing hundred-mile-an-hour winds now racing toward Australia, following just behind the one that was dying, as most storms in the region died, pressed up against the Great Dividing Range that separated coastal Australia from the desert interior like a wall?

How could the population cope? Being caught by such a weather system when the temperature had been 20 degrees Celsius just a week before was going to be a major disaster.

The government began asking airlines to take tourists home as quickly as possible. Airports across the nation were jammed. It was like a war zone.

This time, the southern horizon turned black. Overnight the storm came in, bringing unseasonable tides with it, and blanketing the whole of New South Wales with ice. Trees, caught in full leaf, some in blossom, were overwhelmed. The awful, explosive crack of their destruction punctu-

ated the howling of wind and rumble of thunder.

Estimates were that twenty percent of the trees within fifty miles of the coast were going to be damaged. Then fifty percent. Then, as the ice storm grew more and more intense, all.

Power lines, weighted down by ice, fell one after another. The whole regional power grid was soon down, and repair crews were completely incapable of coping with either the weather or the thousands of breaks.

Large parts of the region would be without power for months, maybe even longer. Meanwhile, temperatures kept dropping. Makeshift emergency shelters were set up in schools, in factories, anywhere a large enough sheltered area could be found. Emergency response efforts were chaotic because the disaster had been so unexpected.

Wind howled down the streets of Sydney, through the lovely parks, and up a steel-gray harbor now choked with mean waves. Here and there, a ship struggled along, and on the empty gray horizon, more distressed vessels could be seen hanging offshore, awaiting word that it was safe to make harbor.

Among those ships was the Seaborne Master, *a massive supertanker. Its last report was similar to that of the* Exxon Invincible: ". . . am breaking up in extreme seas."

From the Russian Federation to France, from Japan to the United States, meteorologists and climatologists were exchanging increasingly frantic E-mails: "What's happening?" "What's going on?"

Nobody realized that events in the Southern Ocean and the North Sea could be linked by a single earthly cause. The only explanation that seemed logical was that there must have been some sort of change in the output of the sun. NASA was called upon to use its solar orbiting satellites to detect what it was.

Then Typhoon Max, which had been building strength in a relatively unpopulated region of the Pacific, suddenly changed direction and headed for Japan. The world forgot about arguments between

NASA and the U.S. National Severe Storms Laboratory over whether or not the sun had suddenly gotten cooler. What would happen when a storm bearing sustained winds of 221 miles an hour struck Tokyo?

An entire species held its breath.

4

Mankind the Unknown

~~~~~~~~~~~~~~~~~~~~~~~~

THE HUMAN SPECIES POSSESSES AMONG ITS EAR-
liest myths an extensive record of some sort of upheaval. Different cultures describe it differently,
but the one feature that is almost universal involves violent weather. For the most part, the stories concern flooding. If we can accept the idea
that our ancestors may have been capable observers of their world, then these myths may turn
out to be carriers of information, possibly urgent
information.

Throughout the history of our species, from
before we were even fully human, the earth has
been experiencing a massive and violent weather
cycle. It has caused our climate to fluctuate
between long periods of horrendous cold and

short ones of benign warmth, which end abruptly and probably violently.

Mankind has remained in southern and central latitudes during most of this time, migrating slowly from Africa into Asia, generally avoiding more northern habitats. Only during the closing millennia of the last ice age, it is believed, did expansion into these regions take place, when the Cro-Magnon peoples began to move north, supported by stone tools of really sophisticated design, better clothing, and more effective shelter building. This amounted to an explosion of human culture and technology on a scale never seen before.

The assumption has always been that there was a steady, linear progression of development after that and that gradual change adequately describes the growth of human civilization.

But the record does not necessarily support this view. What the record supports is a less Victorian, more modern view that cultural growth is much like physical evolution, with long periods of equilibrium being punctuated by sudden bursts of change. It also supports the idea that even very advanced civilizations can and do die and become lost.

As we will see, it is perfectly possible that there was an advanced civilization on this planet that predated our own and that many of our most ancient myths are actually devalued versions of that civilization's own perfectly accurate but very different descriptions of the world in which it found itself and of the disaster that killed it.

There is even evidence that it understood what was destroying it and left us a warning.

When civilizations become extinct, their achievements can be lost, even for thousands of years . . . even forever. The fragility of civilization is most recently evident in the collapse of the Roman Empire.

A highly sophisticated, technologically potent, and economically efficient civilization had spread throughout Europe by 200 A.D. It had a universal language, a universal currency, and a single government that operated according to a consistent set of written laws. It was literate and healthy, highly organized and durable. It was organized around a system of roads that in many places remain in use to this day. A combination of adverse changes eventually caused it to fall, whereupon what would seem to be an unlikely aftereffect occurred.

The Roman Empire was lost to the memory of the common man. Because we know it so well now, we can hardly imagine how complete this process actually was. Within three hundred years of its demise, western Europe had degenerated into a welter of tribal fiefdoms speaking dozens of languages, where writing was almost a lost art and money was considered a form of magic. Roads were abandoned. Paris and London, which had been thriving cities under the Romans, dwindled to besieged villages. Even so, they were among the largest cities in Europe. Shepherds grazed their flocks in the hollow ruins of Rome itself. The rule of law was abandoned in favor of the rule of the weapon. Rome's glory became a myth. It would not be until the Italian Renaissance, a thousand years later, that the glory of Rome would begin to be recovered. Of all its magnificently conceived institutions, only the Catholic Church survived its fall. It would not be until nearly two thousand years later that Europe would again have a single currency and be moving slowly toward an effective central government. There is no indication that Europe will ever return to a single language.

If such a potent civilization could be lost within historical times, what about more distant eras? A sufficiently violent upheaval might leave little more than myths and mysterious ruins, which was all that remained of Rome by 900 A.D.

The world is, in fact, filled with strange ruins, the best known example of which is the Sphinx. It has been thought by archaeologists that it was built by the Pharaoh Khafre in about 2500 B.C. This is because the symbol of Khafre can be seen on a stela erected in front of it by Thutmosis IV when he restored it.

There are three problems with this theory. The first is that there is a likely reference to the Sphinx on another stone record, the Inventory Stela that was discovered by Egyptologist Auguste Mariette in the nineteenth century. This stela was created before the reign of Khafre, so it is hard to see how the Sphinx could be attributed to him.

The second problem is that the Sphinx has been carved out of a huge piece of sandstone, much of which is below the level of the surrounding desert. This means that it is constantly being covered by sand. Who would build a monument in a place where it was destined to be buried within a matter of a few years?

The third problem is the largest one: geologists have made a powerful case that the Sphinx was eroded by water. But there has not been enough rain in the Egyptian desert to account for this erosion. Scholar John Anthony West argued that not only the Sphinx but the enigmatic building known as the Osireion as well showed scientifically irrefutable evidence of such erosion. His conclusions have been endorsed by three hundred members of the Geological Society of America. Their findings have been rejected by Egyptologists, largely because they have no way of accepting them, given what is known about the state of Egyptian culture during the time that the structures are claimed to have been built.

If the Sphinx was eroded by water, then it had to have been constructed *before* historical times, by a lost civilization. This must have happened at least ten thousand years ago, when Egypt had a much wetter climate. This is the stark and inescapable reality. But excavations show that the only people living in Egypt at that time were primitive hunter-gatherers.

So how could they have created a magnificent structure like the Sphinx? It is enormous, a city block long and six stories high. To carve it today would take hundreds of skilled stonemasons decades. And yet the only tools we find in Egypt at that time are simple ones. They were making pottery, weaving, and hunting with stone-tipped spears and bows and arrows.

To see the Sphinx, to actually stand before it, is to be in the presence of a magnificent civilization, not a sparse population of hunter-gatherers. So Egyptologists are understandably left helpless in the face of the geologists' claims. How can they agree to a dating that moves such an incredible engineering achievement back to a time when pottery making was a recent discovery and the primary carving tool was flint?

The world is littered with other such artifacts. Could it be that a civilization of some substantial power existed in some isolated or lost area and created monuments worldwide? Could it be that it was destroyed by something so terrible that only its most massive works remain?

If so, then why aren't the unexplained structures stylistically consistent? In truth, they are vastly different in appearance, claiming only one consistent feature: they are massive and fantastically well engineered, and we have no idea who made them or how it was done.

In rejecting the idea that these monuments are as ancient as they would seem to be, science claims that mankind simply hasn't been competent in things like writing and language long enough to have created two fully developed worldwide civilizations. But there is some very surprising evidence appearing right now that might force a change in this view.

The first signs of the presence of language among human beings have been thought to be about forty thousand years old, when complex activities like cave painting and trade in stone artifacts begins. The evolution of a neck long enough to support complex speech appears to have taken place less than a hundred thousand years ago. It has been assumed that complex culture requires complex speech, just as civilization requires writing.

But now there is evidence that not only were we capable of speech much, much longer ago—as much as four hundred thousand years ago—but technologically based activities were occurring even longer ago than that.

In 1998, scientists at Duke University announced that they had discovered that hominids living four hundred thousand years ago had the full modern complement of nerves needed to speak in the same way that we do. Previously, it had been assumed that not even Neanderthals could speak, that it was a recently acquired attribute—in fact, the main thing that made Cro-Magnon man so much more capable than his ancestors.

The ability to speak and the ability to utter long words quickly enough to have a complex language, though, are two different things. Human babies utter only simple sounds because at that age the larynx is still low in the throat. As the child matures, the larynx ascends. This happens slowly, though, and the speech of a ten-year-old is about half as fast as that of an adult.

Probably, hominid speech was simple, made up of the sort of words that a two- or three-year-old is now capable of forming. Language studies bear out this assumption. They reveal that the oldest human words are mostly structured around the kind of *e* and *o* sounds that babies can utter. Nevertheless, early hominids must have been capable of some level of speech, because there is new evidence that they did things that require it.

A group of scientists at the University of New England in Australia have discovered strong evidence that a human ancestor called *Homo erectus,* which we had dismissed as little more than a clever ape, was actually making journeys by sea—out of sight of land—nearly a million years ago. Language of some kind would have been essential to boat building and navigation.

The reason that they came to the conclusion that open water navigation was taking place then is that stone tools have been found on the island of Flores in Indonesia that were buried in volcanic ash that is over eight hundred thousand years old. Fossil plants and animals found near these tools dated from the same period. So this is a very positive dating.

To get to Flores from Java at that time, where we know *Homo erectus* was living, it would have been necessary to cross straits between three islands. Each one was about fifteen miles wide. They not only crossed the straits, but established a community on Flores, so dozens of them must have been involved.

Considering that these were prehuman creatures, it seems incredible that they could do this and suggests yet again that we need to fundamentally revise our ideas about the actual intelligence and abilities of prehuman beings. And also, perhaps, to set a much earlier date for the evolution of the ability to create civilizations.

We have always assumed that civilization—that is to say, ordered society—was dependent on the written word. It may be, however, that a very complicated social organization that included things like engineering ability evolved first, and that writing—like complex language—was an aftereffect.

The date that civilization evolved will never be found to be as far back as a hundred thousand years. But it might be a lot farther back than seven thousand years. If language, boat building, and navigational skills were present so much earlier than previously thought, the suggestion is strong that civilization, also, might have begun before we previously assumed, especially if early man was as much more intelligent as recent discoveries suggest.

In 1992, German paleontologists were frantically studying deep layers of sediment before they were due to be excavated by a strip-mining operation. Suddenly, on October 20, they located something incredible: an intact wooden throwing stick over four hundred thousand years old. Such sticks represent a technological advance over the lance, in that they offer throwing leverage that increases speed and distance. But the implications go further than that: To use such a tool, one must understand the effect of wind and compensate for it, distance must be correctly judged, and it takes considerable skill—skill that must be taught and practiced—to throw accurately with a lever. In other words, there must be a developed intelligence and an organized community to enable its use, let alone make and preserve the discovery in the first place.

Then a real stunner of a discovery was made. Not only did these people use throwing sticks hundreds of thousands of years ago, they made spears so sophisticated that they are similar to modern Olympic javelins. They are more sophisticated than many spears made in historic times. They were carved out of spruce trees, with the oldest and densest part of the wood at the bottom of the tree used to make the head of the spear. Doing it this way was harder, because it meant that the thickest part of the tree, the base, had to be carved to a point with stone tools we know to have been crude. Remember that these creatures were not fully evolved human beings.

Their intelligent and sophisticated manufacturing technique meant that the spear could be thrown farther, faster, and more accurately, because the heaviest wood was at the tip, giving it much better ballistic properties.

To even realize that this would be useful implies at least a basic understanding of ballistics and an observer's knowledge of trajectories.

Previously, it had been believed that the first signs of complex human thought and action began about a hundred and fifty thousand years ago. But now that view has changed. Complex thought and action began as much as eight hundred thousand years ago, or even longer ago, given that there could be discoveries not yet made that push the date farther back.

We were thinking, planning creatures making sophisticated tools even before we were fully human.

And yet, in 1998 it was discovered that the brain capacities of early hominids were smaller than had previously been estimated. Using a CAT scanner, scientists for the first time obtained accurate estimates of the cranial capacity of a creature thought to be an early human ancestor, *Australopithecus*. It proved to have a brain about the size of a chimp's, not, as had previously been thought, larger. This suggests that volume measurements of other hominids will have to be reduced as well and that these creatures were capable of complex thought and sophisticated toolmaking even though their brains were small and their ability to speak was limited.

However, recent studies do not support the idea that intelligence is dependent on brain size. Brain structures appear to be more important. For example, an African gray parrot, with a brain the size of the end of our little finger, is considerably smarter than a hippopotamus, whose brain is a hundred times larger.

Even with his smaller brain, early man could talk, sail the seas, and make sophisticated tools, all of which was happening eons before we had previously believed.

Thus the mind is much, much older than we thought. Hundreds of thousands of years ago, people who still looked a good bit like apes were thinking about the ballistic properties of their spears. They had been engaging in navigation that took them out of sight of land already for hundreds of thousands of years more.

But does this also mean that we have been civilized for longer than currently believed? No, but it certainly does mean that we could have been.

Have we built and lost a whole civilization before this one, lost it to some terrible, nameless upheaval that periodically sweeps the world clean of all our works, so that we entirely forget who we were and what we accomplished?

The latest evidence is sobering.

5

A Lost World

~~~~~~~~~~~~~~~~~~~~~~~~~~~~~~~~~~~~~~~

NO MATTER HOW COMPELLING THE EVIDENCE
that huge monoliths like the Sphinx are from the
deep past, the unanswered question has always
been the same: If these things were built so long
ago, then where are the cities of the builders?
Where are their tools? More important, where is
the skilled manpower?

We can estimate the population of Egypt at the
time John Anthony West claims that the Sphinx
must have been built. Judging from the settle-
ment density that has been established along the
Nile, the entire region must have contained fewer
than a hundred thousand souls, widely dispersed
in villages containing fifty to five hundred people
each.

How could such a sparse population, practicing only rudimentary agriculture and not even storing grain, possibly have contributed the thousands of workers that it would have taken to create the Sphinx?

And yet it's there, and the geologic evidence of its age is so compelling that it is almost impossible to refute. The age problem is a straightforward one. The condition of the sandstone from which the Sphinx was carved indicates that it has been eroded by water. The problem with this is that Egypt has been a desert since before the rise of the pharaohs, so where did this water come from? Structures that were unquestionably built by the pharaohs, also out of sandstone, show no such erosion.

So whoever built it must have come to Egypt to do so, then left, taking their tools with them. Either that, or we simply have not yet found their cities. Perhaps they built away from the Nile, in areas that were fertile then but have been empty desert for so long that we have seen no reason to examine them. Something like this must have happened in Brazil, where the remains of a vast area of ancient cultivation exists, marked by hundreds of miles of canals and capable of supporting a population of millions. But there remains almost no record at all of this culture.

There has been a discovery made off the coast of Japan that also suggests that civilization is much older than previously thought. It has given credence to claims that many monoliths and monolithic structures whose dates are in dispute—such as the Osireion, the Sphinx, and the enigmatic Peruvian city of Tiahuanaco—really are as old as observers such as John Anthony West and Graham Hancock have been suggesting.

And now that the Japanese find has been confirmed, another archaeological site discovered in the South Pacific many years ago may also begin to make sense.

Seventy-five miles southwest of Okinawa lies the little island of Yonaguni, chiefly known for its small population of Yonaguni ponies, an equine subspecies unique in the world. But something else unique has been found near the island. In 1988, scuba divers discovered what was thought to be a huge natural formation seventy-five feet underwater. But a geologist at Ryukyu University in Okinawa, Dr. Masaki Kimura, investigated the structure and said, "the object has not been manufactured by nature. If that had been the case, one would expect debris from erosion to have collected around the site, but there are no rock fragments there." Additionally, he discovered what appears to be a road around the gigantic structure, which is almost as big as the Great Pyramid at Giza. Robert Schoch, professor of geology at Boston University, dived the site in April of 1998. He said, "It basically looks like a series of huge steps, each about a meter high. Essentially, it's a cliff face like the side of a stepped pyramid." He added, "It's possible that natural water erosion combined with the process of cracked rocks splitting created such a structure, but I haven't come across such processes creating a structure as sharp as this."

The possibility that this is a man-made object is strengthened by the fact that smaller underwater stone mounds have been discovered nearby. These are also made of stepped slabs. They are about thirty feet across and six feet high.

There are no records at all of people capable of building on this scale in prehistoric Japan, although the Jōmon people, who inhabited much of Japan ten thousand years ago, are known to have built large wooden structures. But as we have seen, there are no records of people able to build the Sphinx 10,000 years ago, either.

At least, none that they themselves left. But peoples as diverse as the Babylonians and the Maya have legends that civilizing deities came to them out of the sea. But what about the Okinawans? If this ancient civilization actually existed, surely they would also have a similar memory of it, since they must have been living in what would then have been an area of highlands close to one of its important centers.

There is indeed such a legend on Okinawa: the god Nirai-Kanai came to them from the sea bringing happiness. Like the Sumerian Oanes, the Viracochas of Peru, and Quetzalcoatl of the Maya and the Aztecs, this deity seems to have arisen from the ocean. Strangers coming on boats could have given rise to such a myth.

So, how old *is* the Japanese site? Dr. Teruaki Ishii, professor of geology at Tokyo University, explains that the land on which the structure was built sank into the ocean about ten thousand years ago. This was about the time that the Sphinx was built, if the geologists are correct, and also during the time of a mysterious and violent event that led to the extinction of many different animal species, with a strange concentration on the larger creatures. It was during this time that mammoths, for example, ceased to roam the world.

Of course, the fact that the monolithic structure sank into the sea thousands of years ago is no indication of when it was actually built—except that it must be at least that old.

It would help, of course, if we could determine whether it is in an incomplete state. But, just as there is no erosional debris nearby, there is no evidence of construction, either. So it seems likely that it was built some time before the land actually sank. But how long before? As yet, there is no way to be certain.

If we could compare it to buildings elsewhere in the world known to be of great antiquity, it might help us date it. But the structure has been mapped, and it seems to have no architectural relationship to any other known building, except for the fact that, like other very early structures, it is unadorned. It presents a strange picture indeed.

There are many layers to it, ramps and platforms all built with straight sides. But the overall appearance is chaotic, as if the builders had magnificent construction abilities but very little idea of architectural planning. Unless, of course, the mind that made the object was very different from our own, and what appears to be chaos actually conceals a kind of order that we no longer understand.

Maybe the object is natural, no matter how man-made it appears. If somebody that long ago made anything the size of the Great Pyramid, surely there would also be remains of their civilization elsewhere in the Pacific Basin.

As it turns out, some of the most mysterious remains in the world are in the Pacific Basin. Among these are the stone ruins of Nan Madol on the island of Ponape in Micronesia. They appear to be made of huge logs, but the "logs" are actually basalt.

The ruins are extensive, stretching over ninety man-made islands that cover eleven square miles. The basalt "logs" were dragged many miles and then rafted across bays to create the structures. Some of the blocks weigh fifty tons, and the city contains an astonishing 458,000 tons of basalt.

To first construct nearly a hundred artificial islands, then to add these enormous basalt structures to them would be a huge task even for modern engineers. Even more difficult would be to create the extensive system of underwater tunnels in the area, which have been hewn directly out of the coral reef. How this was done without modern breathing equipment is a mystery.

The fact that much of the city is sunken suggests great age, but no geologic evidence has been gathered to support a date for the sinking.

In the 1960s, the Smithsonian Institution sent an expedition to Nan Madol to gather data. It concluded that the structures were nine hundred years old, based on carbon dating of ashes at the bottom of a fire pit inside one of them. But since there was no way to be sure that the fire was made by the builders of Nan Madol, this conclusion is flawed. It is further challenged by the fact that there appears to have been nobody in the area capable of engineering Nan Madol at that time.

In the 1970s Steve Athens of the Pacific Studies Institute in Hawaii dated some pottery shards found near the structures. Thermoluminescence showed that they are at least two thousand years old.

The date of this pottery was a surprise because when the people of Ponape came into contact with Europeans just a few hundred years ago, they did not know how to make pottery. They did not even have oceangoing canoes, let alone anything remotely capable of moving a fifty-ton block.

So, at the very least, the culture must have regressed dramatically. The local population has legends about Nan Madol, most specifically that the blocks were moved through the air by magic.

How were they actually moved, though? The most logical conclusion would be that it was done with rafts. The sea bottom in the area is littered with basalt logs that apparently sank when rafts were swamped. These structures cannot be age-dated, at least not with present technology. How old they might be is anybody's guess. It would help if they could in some way be connected to an early civilization. But the first signs of human habitation in Micronesia appear about 1500 B.C., suggesting that the older pottery shards must have been the products of a precursor population. Judging from how little of this pottery has been found, this population must have been sparse.

As with the Sphinx and the sunken structure off Yonaguni, it is hard to relate these huge constructions to the primitive peoples who were apparently the only inhabitants in the area in which they are found.

Maybe, however, they weren't the only inhabitants. The bones of people much larger than the native Micronesians have been excavated in the region. Similarly large human bones, incidentally, have also been found in the Americas. In the United States they've been the subject of a dispute between Native Americans and archaeologists. A 9,300-year-old skeleton found in Kennewick, Washington, in July of 1996 appears to have the features of an Asian—something like those of the Ainu people of Japan. The bones have been locked up because the local Umatilla Indians want to rebury them without further study, but their existence is another strong suggestion that the history of man over the past ten thousand years has yet to be completely written.

Another mystery is that there are standing columns at Nan Madol, but they are all under water. There are no standing columns on land. There is no obvious explanation for this. One thing, however, is clear: Nan Madol is a genuine mystery. It is at least two thousand years old, but there is little other evidence even at that time of a civilization in the Pacific that was powerful enough to support such a massive building program. It seems difficult to believe that such a civilization would not have traded with China, for example, which was the world's largest and best organized state at that time. And yet there are no Chinese records of such trade. Aside from pottery shards, there are no artifacts associated with Nan Madol.

Was Nan Madol even a city? Certainly it was a place that required traffic between different sites, thus the tunnels and canals. But why build tunnels in the first place, and why does Nan Madol have more the look of a fortress than a city? The largest construction, Nan Dowas, is surrounded by immense walls.

This area of the Pacific was once very different than it is now, but geologic changes extensive enough to have placed Nan Madol entirely on dry land have not occurred for tens of thousands of years.

Given that Pacific navigation, judging from the finds on the island of Flores, has been a part of human life for nearly a million years, mankind might well have reached the islands much earlier than we currently think. However, there is still a problem with Nan Madol, as there is with the structure off Japan. Nowhere in all this history— not a thousand years ago or three thousand years ago—do we find any evidence of a civilization developed enough to have accomplished these massive feats of engineering and construction.

So, could it be that we are actually looking at remains that are *much* older than anyone has yet guessed?

Possibly. There are some mysterious ruins in the Pacific that have been carbon-dated, with fascinating results. On the island of New Caledonia and the nearby Isle of Pines there are found numerous cylinders made of cement. They are three to six feet in diameter and three to eight feet long. They are made from a very hard lime-mortar cement, which contains pulverized shells.

Since the shells are organic, they are accessible to carbon dating. Radiocarbon dating done by the Centre des Faibles Radioactivités in France in 1966 indicates that the cylinders are between seven thousand years old and thirteen thousand years old—dates that are right in line with the time that the Yonaguni structure sank into the sea.

The cylinders, when they were observed by Andrew Rothovius for the *INFO Journal* in 1967, were found inside gravel mounds that were obviously artificially constructed as well. Four hundred of these tumuli (mounds) had been found on the Isle of Pines and seventeen on New Caledonia. There were no bones, no sign of the use of fire, and no other artifacts around them that might have challenged the dating of the shells. The cylinders were buried vertically in the tumuli, which were themselves eight or nine feet high and about 300 feet in diameter.

Neither the cylinders nor their placement in the mounds suggests a known purpose. In this sense, these ruins are similar to Nan Madol and Yonaguni, which do not look as if they have been built to serve any known purpose, either.

The Isle of Pines ruins look much more like something that might have had some sort of forgotten technological purpose. What that might have been remains completely unknown.

Not only the Pacific, but the whole world is dotted with strange monoliths or monolithic structures that appear to come from a very distant time.

There are other sites of great antiquity and almost unimaginable size. One of them, the platform at Baalbek in Lebanon, probably could not be built today, not without manufacturing special tools, lifting devices, and transporters.

Like most anomalies, the Baalbek structure has been largely ignored in favor of maintaining present theories about the past. When evidence doesn't fit theory, scientific method is in danger of breaking down unless theory is revised to accommodate the evidence. Scientists, being human, have a tendency to fall into the trap of doing the opposite, defending theory by ignoring evidence. In this way, science ends up protecting beliefs rather that explaining facts.

No matter what may be said about it, the awesome strangeness of the platform at Baalbek cannot be explained away. It has been there for thousands of years and is indisputably a constructed object, not a natural one. Nobody knows who built it, although there are numerous later remains in the same area. Traditionally, it was connected with sun worship, but it was put to many other religious uses in its history.

The platform itself consists of three gigantic stones known collectively as the Trilithion. These three stones are among the heaviest objects ever moved by man, possibly the heaviest ever to have been moved as much as thirty-five miles, which is the distance between the platform and the quarry where the stones were cut. A fourth stone remains in the quarry unfinished.

Not only is the platform made of gigantic stones, they are actually raised off the ground to a height of twenty-six feet on a foundation of smaller stones. The original structure was apparently a massive temple or ritual space of some kind. Later cultures built on it, most notably the Romans, who left a spectacular temple to Jupiter.

Each *single stone* in the Trilithion weighs a breathtaking six hundred tons. It is hard to communicate just how big that is. The stones look like something from some other, larger reality. They are each more than sixty feet long. All three are thirteen feet high and ten feet across. Perhaps it is easiest to compare them in weight to something familiar to us. One of them weighs as much as twelve hundred average passenger cars.

It is possible to conceive of moving such an object. We can dry-dock ships weighing tens of thousands of tons. But moving something so large across miles and miles of open country is another matter entirely. Given the enormity of the task, one would think that there would be some evidence remaining of the work that had been done to move the stones—a causeway, perhaps, between the platform and the quarry, or some indication on the stones themselves that they had been dragged.

But there is no such evidence. And given our understanding of the materials available to the peoples in the area during antiquity, it is not possible to see how they could have moved the stones at all. The ancient world had no ropes strong enough to bear the tension necessary to drag such objects, nor wheels or rollers able to support them without crumbling. If they had made huge rollers out of a stone like basalt capable of bearing the weight involved, then where are they now? Had rollers been made of the same limestone that was cut to make the slabs, the places where they had been quarried would also have been found. It is doubtful, however, that limestone rollers large enough to bear the weight of the stones without crumbling could even have been rolled. They would have been almost as heavy as the slabs themselves.

Even if the builders had managed to get the stones onto rollers, then it would have taken twenty thousand people moving at a rate of about a tenth of a mile a day to transport them. Given that their collective strength could not be distributed with a rope, they would have had to push the stones along on their rollers by hand. But there isn't enough surface area available even on the longest faces of the stones to enable the number of people who would be needed to move it to press against it all at the same time.

The platform at Baalbek could not have been built using the tools and materials available to the cultures known to have been in the areas where it is found. And as far as the ruins off the Japanese coast and elsewhere in the Pacific are concerned, we once again have not found the faintest trace of any civilization that could have built them, let alone the engineering methods that were used.

There is even evidence that historical civilizations may have used tools and methods of which they left no record.

While it has been theorized that the Great Pyramid at Giza, for example, must have been built using a series of earthen causeways to raise the stones higher and higher, the earth that would have been used has never been found. And this is not a small amount of dirt: it would have covered an area larger than the pyramid itself. It would have been quarried, but no such quarry exists. In addition, there is a sarcophagus inside the Great Pyramid that has been made out of a solid piece of granite. It must have been cut either with bronze saws studded with diamonds or saws of a much harder metal. But diamonds have never been found in Egypt, and at the time the sarcophagus was supposedly made, metals harder than bronze were unknown to the Egyptians or anybody else on earth.

Not only that, the sarcophagus is cut to tolerances that have only recently been attained with modern tools. The volume of the interior is 1166.4 liters. The exterior volume is 2332.8 liters, precisely twice as much, measured to the tenths.

How could craftsmen using primitive equipment such as the Egyptians supposedly possessed possibly have done this? We don't know.

It gets stranger. The sarcophagus was hollowed out by a process of coring. We know this because the cores have been found in the quarry where it was made. According to Egyptologist Sir Flinders Petrie, the cores were drilled using extraordinary pressure, in excess of a ton of weight on the drill. And one core, which shows a one-inch angle of declension, would have been created under truly extraordinary press pressure, in excess of three tons.

How the Egyptians could have achieved a drill-press weight of a ton, let alone three or four tons, is unknown. More than that, no drill made of bronze could have withstood the pressure. It would have been crushed long before it even began to turn.

The Egyptians have also left things like hieroglyphs carved into hard stone such as diorite and quartz that must have involved the use of extremely fine tools. To conceive of a bronze or even a steel drill bit that could have sustained the pressures needed to drill out the sarcophagus in the King's Chamber is difficult. But it is even more difficult to see how they could have made a bit fine enough to do the delicately carved hieroglyphics found on diorite bowls, many of them dating almost to prehistoric times.

The lines carved are 1/150 inch wide and must have been made by something harder than quartz that had an edge only 1/200 inch across. Author Graham Hancock, traveling through Egypt doing research for his book *Fingerprints of the Gods,* found that "more than 30,000 such vessels had been found in the chambers beneath the Third Dynasty Step Pyramid of Zoser at Saqqara."

Hancock, formerly an engineer, considered the tool that must have been involved in some of this work "almost unimaginable." This was because some of the vessels were long and thin, with narrow necks and "often with fully hollowed-out interiors." Vials so tiny that they are almost microscopic have also been found. Hancock continues, "what was really perplexing was the precision with which the interiors and exteriors of these vessels had been made to correspond."

The fact is that the ancient Egyptians had no known tools that were capable of creating these vessels. We know this because modern stone carvers using tungsten carbide drills couldn't do it.

So what's the answer to the enigma?

Well, for one thing, the enigma doesn't stop in Egypt. To see a little more of the wonders created by our forefathers, let's go to Peru.

There are many extraordinary ruins here, most of them casually attributed to the Incas. But the way that they were actually built has never been explained—at least, not in a way that doesn't involve some sort of magic. But these ruins don't suggest magic. Like the other remains scattered around the world, they suggest some kind of remarkable lost technology.

This technology included the ability to manipulate great weights and to create and use tools of extraordinary power and fineness. In some respects, it was actually beyond the achievements of modern science.

The single most incredible achievement in the known history of stone transport involves the ancient fortress of Sacsayhuaman. This astonishing ruin, lying north of the ancient Inca capital of Cuzco, may well be the most accomplished piece of stone engineering on the planet. This is not because of its size. The stones involved, at three hundred tons each, are only half as big as those at Baalbek. But they are shaped to a tolerance so fine that a hair cannot be inserted into the joints between them. More than that, the engineering achievement involved in moving these stones up through the mountain passes is more than incredible. It borders on the miraculous.

In his sixteenth-century chronicle *Royal Commentaries of the Incas,* Garcilaso de la Vega, whose mother was allegedly of royal Inca blood, said that an Inca king had tried to move a gigantic boulder from a few miles away to add to the fortress. It was hauled by hand by over twenty thousand men, who eventually dropped it, crushing three thousand to death.

In other words, the Incas as the Spanish found them did not know how to build some of the buildings that they claimed as their own. Certainly, they used them, and for this reason archaeologists simply assume that they also built them. But no explanation was ever forthcoming from the Incas about how they might have done it.

The world may be littered with the debris of a much earlier civilization than any we have identified. How much earlier we don't know.

What could have happened to destroy people capable of building on a scale that humbles even the engineers of today and whose tools for detailed work reduce modern designers to amazed silence?

Whatever destroyed this civilization, it spread across the earth in the twinkling of an eye, bringing with it a destructive fury of unprecedented horror. But the world that it destroyed apparently was not entirely silent on its nature or its potential to return. The people of that world left us a message about it—in fact, many messages. When we look past our own comfortable assumptions about those messages—that they are nothing but meaningless myths and legends, the confused output of people who understood their world not at all—we will find an amazingly coherent structure, one that seems designed to give anybody who understands only the most basic laws of nature access to their meaning.

When we unlock that meaning, we will discover what happened to that old, lost world—and why so long ago it pointed its finger straight at our era, at the next few years, at us.

6

The Tokyo Express

The Japanese Meteorological Agency was in
an emergency situation. Supertyphoon Max
was going to come inland, and the bureau
could not get accurate readings of its wind
speeds because it was destroying all the measuring instruments in its path.

The wind speed data coming from Japan's
Advanced Earth Observing Satellite seemed
impossible to believe, but there was no alter-
native. The nation had somehow to prepare
for a post–Category 5 supertyphoon, a

storm bearing sustained winds far in excess of 155 miles per hour. What they were seeing were sustained winds reaching 213 miles per hour for long periods of time.

The storm had the tight, well-formed configuration of all truly powerful typhoons. It wasn't like most of the typhoons that struck the Home Islands, their windspeeds diminished by previous brushes with Taiwan or the Philippines. Normally, the greatest danger Japan faced was from flooding. But this storm was different; it had lingered in the central Pacific without touching any major landmasses.

It was Japan's worst nightmare, a pristine typhoon hitting the islands with full force. The country began to prepare. But what could be done? It was going to make landfall in just a few days, and there was no way to evacuate a coast populated by fifty million people. Where would they go? Japan had to face Max on the beaches and in the harbors.

Storm surges of thirty feet seemed possible. Millions of people were going to drown.

To understand why so little could be done even with nearly a week's warning, it is necessary to go to Tokyo, to get a feeling for the city, to comprehend its vastness and its complexity, to come to know the fragility of its essential systems—a fragility that it shares with every other highly organized, technologically advanced city in the world.

To call Tokyo beautiful is an understatement. It is a soul-stirring combination of the delicate and the awesome. It is without question the most intricate human habitation ever created. Without named streets, it is a city that demands to be known. But Tokyo is also a hidden city, a place that hoards the very secrets the observer needs. The stranger can wander the streets for hours with his landmark in sight, never getting any closer despite the careful directions obtainable from any local on any sidewalk.

Tokyo is truly an ancient place. It was already a thriving metropolis when London was still clawing its way out of the Dark Ages and Paris was just beginning to relight the candles that had died with the fall of Rome.

Tokyo: the Ginza of tourist fame, blasting with high energy, high style, and high prices. The jam-packed, surprisingly efficient trains. The surreal traffic. Tokyo, where the most perfect fruits the world has to offer are sold from spotless stands, at prices that make Westerners gasp. Tokyo, where violence is rare—except in theaters, where it is incredible—where sex is not connected with sin—not necessarily, that is. In Japan, the pleasures of the flesh are just that—pleasures. Or so it appears on the surface.

Tokyo is an ancient place, but not an old one. During World War II, it was firebombed almost completely to ashes. Aside from the grounds of the Imperial Palace, there are few historic structures. However, there is a sense of history here, the feeling that this place is inhabited by an ancient people. As the tourist leaves the center of the city, moving into side streets, a subtle world opens its door just a little. Leave the Ginza, and Tokyo's other districts unfold themselves:

Ueno and Asakusa, Tsukiji and Shinjuku. Leave the center of the city, and the streets quickly narrow. Uniformed children pass; shops vend everything from gold chains to fresh-cut sushi; crowds and crowds of people surge by.

One wonders about the true meaning of this place, the true nature of its past . . . wonders about the dreaming ruin in the waters off Yonaguni and the strange fact that Fuji-san, a volcano, has been chosen as the national symbol of tranquillity. Is it hope that identifies Mount Fuji with peace, or a desire to appease the restless earth?

Much of Tokyo is built of glass and aluminum and concrete, but much more is still constructed of far more delicate materials. The Japanese may live in apartments that look entirely Western, but they also live, in surprising numbers, in the same sort of lightly built structures that once defined the community.

Typhoon Max moved with the nasty grace of a snake. It would swell toward land, then cringe back into its mid-Pacific lair, would *rush an island, then seem to rest . . . or lie in wait for shipping.*

During its days in the mid-Pacific, it dispatched no fewer than nine major ships and fourteen smaller ones. In each case, the vessel thought itself safe. In each case, the chaos that was breaking out in earth's atmosphere caused the storm to make an erratic lunge, and another ship radioed the despairing message, ". . . am breaking up in extreme seas."

Rule one: A storm will continue until the imbalance that spawned it is corrected. As long as a typhoon sits over warm water and the upper atmosphere remains cold, it will survive.

Max had started in the same region of the Pacific as Typhoon Stella, which had made landfall in Japan on September 16, 1998, killing twenty-three people despite maximum sustained winds of just sixty-five miles an hour.

For Japan, storms starting in this area are extremely dangerous if they turn north-northwest, as Stella and Max both had. This is because they gain strength over the vast open sea, water that doesn't stop until the beaches and cliffs of Honshu and Okinawa.

Normally, however, the waters are much colder in this area, their temperature governed by the flow of the Pacific Circulation. So storms weaken as they approach the Home Islands.

Not this time. This time, there was a major problem with the Pacific Circulation, a problem that had its roots half a world away, a problem that nobody understood. Because of it, the Japanese Meteorological Agency made an error of historic proportions. It predicted that Max would weaken as it approached Japan and issued flood warnings. They decided that the satellite wind-speed data could not be correct.

Just after noon on a dense, sweaty Thursday, Max struck. Massive flooding and extensive damage were expected. Shelters had been set up capable of housing a million refugees.

On the first day, events unfolded much as predicted. Rain fell at a rate of ten inches an

*hour in many places. Streets flooded. A
thousand people were evacuated when dikes
failed along the Arakawa River.*

*The first sign of extreme danger came
when a small coastal town in Shizukoa pre-
fecture reported that waves were battering
the city hall, then went off the emergency
radio network. Meteorologists quickly realized
that the waves involved must have been
nearly forty feet high.*

*The satellite data was correct: This was
indeed the most terrible typhoon ever recorded.*

And yet it was so late in the season.

*And yet typhoons did not grow when they
were over cold water.*

*So the water could not be cold, not even
as cold as it had been a month ago.
Something had changed the surface temperature of the western Pacific—and done it very
suddenly.*

*The winds were horrific. They tore into
the whole nation, devouring houses, sending
cars and trucks flying through the air, strip-
ping aluminum panels off skyscrapers, and
turning their windows into billions of tiny,*

lethal daggers. All over the nation, people clustered around portable radios.

The broadcasts soon failed, though, as one radio tower after another was torn from its mooring and sent tumbling off into the sky. Television transmissions stopped, even cable-based ones. Satellite uplinks were being destroyed throughout the islands.

The power grid collapsed just after midnight of the first night.

And then—only then—did the storm come ashore. What had been experienced so far had only been a prelude, a softening up. Water surged up Tokyo Bay, and residents were horrified to discover that the ocean was filling the streets.

The noise of the water, the roar of falling buildings, the mad howl of the wind—an insane, vicious sound, the shriek of a bloodthirsty ghost—was so loud that voices were drowned out even deep in the strongest buildings.

Thousands were forced into the streets by rising water, only to be carried away by the floods or battered to death by the wind.

And it went on and on and on—until, as all storms do, it ended.

Supertyphoon Max: highest sustained wind speed, 218 miles per hour; death toll, 1,288,704. The most deadly storm in history—so far.

It was only a prelude, like heat lightning, like distant thunder. Paris and London and Berlin, Stockholm and Brussels—all looked to Tokyo and breathed a silent prayer. New York did, too. At least nothing that terrible could ever happen in New York.

Or so they thought: The superstorm was yet to come.

7

The Last Disaster

~~~~~~~~~~~~~~~~~~~~~~~~~~~~~~~~~~

IF THEY HAPPEN, SUPERSTORMS ARE RARE. WE
will discuss the possible mechanism of their formation in detail later, but at present it is sufficient
to say that there may exist in the fossil record an
echo of one that took place between seven and ten
thousand years ago.

About eighteen thousand years ago, the last ice
age reached its maximum. Huge glaciers stretched
as far south as the central United States, and winter
temperatures in Texas were like those presently
experienced in Canada. Sea levels were much lower
than now, with the land line stretching many
miles from present shores.

The world was populated by thousands of large
animal species—the mastodon and the mammoth,

the cave bear and dire wolf, the giant beaver and the short-faced skunk.

For the next eight thousand years, the ice retreated steadily. Eventually, the retreat became a rout. This warming was apparently caused by a persistent increase in methane in the atmosphere. Although it dissipates quickly if it isn't replenished, methane is a potent greenhouse gas, and its buildup at this time led to a global warming event like the one we are experiencing now. There followed, around eighty-two hundred years ago, a period of sharp cooling that lasted approximately two hundred years, apparently caused by a flood of fresh water into the Arctic Ocean.

Did the sudden drop in ocean salinity caused by this flood of fresh water trigger the southward movement of the North Atlantic current of the type that we believe triggers the superstorm? To determine this, it will take a careful reexamination of the fossil evidence. Nevertheless, as will be seen, it probably should be done, because that evidence is disquieting and strange.

Entire populations of animals disappeared, almost exclusively large ones. Eight different genera of mammals ceased to exist in North America. Worldwide, twenty-seven were rendered completely extinct. This process started before the cooling event, but the last extinctions took place then.

It has been claimed that the extinction of large animals in the Americas coincides with the spread of man, a comfortable theory that has perhaps been used to obscure a very thorny and uncomfortable problem, for North America was not the only place where large animals became extinct during this period. There were parallel extinctions all over the world—in Australia, in Europe, in South America. How could a human population that may have also been dropping worldwide due to the weather changes be responsible for even part of this wave of extinctions?

Something much more bizarre appears to have happened, something involving the climate.

The remains of mammoths are found all around the Arctic Circle with food still in their mouths and stomachs that indicates that they were grazing in a temperate climate when they died. Trees have been found that were frozen solid while in blossom. Whatever happened, the event took place on a summer day, which suddenly got much, much colder.

There were millions of mammoths in northern Siberia at the time. One of them, the Adams Mammoth, which was discovered in 1789, was found to be well fatted; another, the Berezovka Mammoth, had fragments of flowers in its stomach. This confirms that northern Siberia was warm at the time, and it suggests that not only must the death of the animals have been sudden, but the bodies must have cooled very rapidly, or these delicate plants would have been dissolved in their stomach acid. It is thought that a cooling to around –150 degrees Fahrenheit, from a starting temperature of around +80 degrees, would have been necessary to stop the digestion process in time.

The soft parts of large animals have been found frozen throughout a vast circle that covers two continents—all above the Arctic Circle, all in states that suggest sudden death by freezing, all in a condition that confirms that their lives were being lived out in a temperate climate. Not only that, masses of mammoth and rhinoceros bones are found heaped on the highest points of plateaus. A total of 116 specific concentrations of animals that died under these conditions have been excavated. Why were they like this? Were they fleeing the floods that occurred as the snow and ice left by a superstorm rapidly melted—floods that are commemorated to this day in the legends of tribal peoples who live to the south of this area in Canada and the Pacific Northwest?

The presence of food in the mouths of the animals suggests that their death was very sudden.

A creature her discoverers christened Dima, a baby mammoth found entombed in ice, had its lungs and respiratory tract full of silt, clay, and gravel, as if it had been suffocated by an extraordinary mass of soil debris.

The vegetative matter found in the animals suggests that death took place in the summer—exactly the time of year a superstorm would be most likely to occur.

The question of how these animals died is controversial. Among the theories that have been put forth there is one that they drowned in very cold lakes and were preserved by the low temperature of the water. If this is so, why did they continue to eat as they drowned, and why, if the lakes were so cold, was the surrounding climate so temperate, as is proved by the contents of the creatures' mouths and intestines and the fossil record itself?

It has also been theorized that the ones that are found standing died when they became trapped after the top layer of tundra melted in the summer and they went through. Their still-standing bodies were frozen in winter and thus found that way.

This theory does not explain why they would not have thawed again the following summer or become the victims of carrion eaters during their long exposure.

What appears to have happened is that the climate went into a state of greenhouse warming due to excessive methane in the atmosphere and these animals died during the extremely violent weather that came about when there was a brief snap back to a cooling trend.

Did this event also end a human civilization? It certainly caused a dramatic drop in human population in the Americas, in Australia, in Europe—although there is, of course, not the slightest trace of evidence of an engineering civilization in Europe or anywhere else at the time. If a change in worldwide ocean currents occurred, these three areas would have been hardest hit.

Whatever happened, it was extremely violent, far more violent than the combination of disease, economic problems, and invasion that ended the Roman Empire. And we have already seen how much Rome's survivors lost, especially those far from Rome itself, and how long it took for the memory to be recovered. Other civilizations, such as that of the Hittites in the Middle East, were forgotten so completely that they were altogether lost to the historical record for thousands of years.

It is possible for history to lose track of an entire civilization, so what happened 7,000 to 10,000 years ago might well have erased the memory of a civilization, especially if its population and records were centralized in an area that was completely devastated.

If such a civilization existed, then it would probably have left some of its ideas behind, which would have resurfaced in the form of myths and been misunderstood by those too ignorant to see its science as anything but magic.

The same thing happened in Rome's case. In the Western world, by the tenth century, the use of money was no longer understood, despite the fact that long ago Rome had even developed a basic concept of credit. When money was reintroduced to Europe from the Arab world via the maritime republics of Italy, it was at first regarded as a form of magic.

One thing that was almost universal among our earliest civilizations was an interest in calendars. Most of these were seasonal and used in agriculture.

There is one universal calendar, however, that is different and has a much more obscure purpose. This is the zodiac. It is a stellar calendar that is designed to measure the angle at which the earth is facing the sun over the twenty-four-thousand-year precession of the equinoxes. It does this by marking the passage of zodiacal constellations through true north. Not only that, it is constructed as a mnemonic device. As long as you can remember the constellations, you need no equipment other than your eyes to place yourself relatively exactly in this vast space of time.

But why? Who cares about a twenty-four-thousand-year cycle? Why even design a zodiac? And yet it is among the most universal of calendars. From culture to culture, the names and configurations of the constellations change, but the purpose of the calendar never does: from time immemorial, it would seem, mankind has retained the strange, apparently pointless ability to measure this huge cycle of time.

In their book *Hamlet's Mill,* Giorgio de Santillana and Hertha von Dechend argue that there is a profound coherence behind many thousands of myths, legends, and calendrical markers throughout the world that point to the notion that the zodiac is meant to identify widely spaced times of upheaval. Hamlet's mill is named after the legendary Danish king and Shakespearean character Hamlet, who ruled wisely and well except for brief periods during which he went mad and devastated his whole realm. Normally, he turns his mill with great regularity. But sometimes, he goes mad.

Whatever its cause, ancient stories suggest that a great catastrophe took place around the time that the last superstorm may have occurred. As Graham Hancock explains in *Fingerprints of the Gods,* "the Mayan *Popul Vuh* [the sacred book of the ancient Quiche Maya of Mexico and Guatemala] associates the flood with 'much hail, black rain and mist, and indescribable cold.'" He reports a legend from Tierra del Fuego in southern Argentine that "the sun and the moon 'fell from the sky.'" According to John Bierhorst in *The Mythology of Mexico and Central America,* there is a Mayan account that states, "It happened that the sun was still bright and clear. Then, at midday, it got dark." There followed years of darkness and agonizing cold.

Such a storm would cause black skies, rain, floods, and precipitate global cooling. We have the evidence of sudden extinctions at that time and the legends of floods. We have evidence of a sudden reversion to a much colder climate at that time that conceivably mirrors the present threat.

Through most of the earth's history, the Tropics have extended well north and south of their present limits, and climate change has been slow. In general, the polar caps have been small or even nonexistent.

The state that we have been in for the past three million years, with recurrent periods of glaciation and thaw, is almost unprecedented in earth's geologic history. The rising of the Central American land bridge not only changed the climate, it forced it into a huge cycle—a cycle that our ancestors may have discovered just before it destroyed them and that they attempted to warn us about.

8

The War Hypothesis

WHETHER THERE WAS AN EARLIER CIVILIZATION
is not central to our argument that we are in danger of a massive shift in climate. But we are exploring the issue at length for two reasons: First, if
this civilization existed, it died during a period of
similar climate change, and we need to know as
much as we can about how and why that happened. Second, it is possible that we can read a
record of its knowledge and experience in the
myths that it left behind and from that gain insights that will help us navigate the dangers of the
immediate future.

We have seen that there is a lack of records prov-
ing the existence of such a civilization. Maybe
that's because it didn't exist, and maybe the con-

ventional explanations for the ruins we discussed earlier are the true ones. Even so, there is the chance that we may gain some useful knowledge by exploring the other possibility a little more deeply.

If the civilization did exist, then where are its artifacts? Was the natural catastrophe that befell it so extreme that they were all destroyed? That does not fit very well with the fact that the unexplained ruins are found all over the world. This civilization might have been highly centralized in an area that was completely destroyed, but if it left structures all across the planet, why not debris as well? Surely there would be some dropped tool, some lost bit of jewelry or sculpture, that would reveal its existence.

Or would there? What if a horrific war took place in the context of violent changes in climate, leaving most of the civilization shattered by the storm and the rest blown to pieces?

If we had an all-out nuclear war right now, three-fifths of our economic infrastructure would be destroyed. We might lose as much as ninety percent of our population. The only people left alive would be the ones living in what are now peripheral areas—backwaters irrelevant to any of the combatants—and the few places that remained untouched by the natural disasters. If only widely scattered pockets of human life remained, then what would be left of civilization? How long would the memory last?

Judging from historical experience, not long.

What would persist would be a folk memory of the fearful power of the disaster. As we lost touch with science, it would begin to be remembered as magic. We would hunger to use its powers, which in memory would come to seem greater than they actually were. Jets might become spacecraft; the Internet, a magical wellspring of all knowledge; the actions of scientists, the rituals of priests. The objects of technology, no longer understood, would become objects of worship.

Anthropologists have observed a similar process at work in the modern day. During World War II, Western technological civilization reached isolated parts of New Guinea for the first time. The local people were awed by the airplanes they saw landing on jungle airfields cleared by the U.S. Air Force. They longed to have access to the wealth of goods they saw pouring out of these airplanes.

Their response was to try to influence the airplanes, replacing the technological knowledge of the Westerners who controlled them with magical versions of their own imagining.

Having no idea of what an engine was or how a wing worked, they assumed that the planes were themselves living beings and that they gave rewards to those who served them well.

These people thought that the machines of the West were gods and that the commonplace actions of the people operating them were priestly ritual.

They tried to attract the flying gods by creating "airfields" of their own and populating them with bamboo "airplanes" and "refrigerators," then moving around them in ritual dances meant to emulate the airmen. Some cargo cultists, as they were called, even worshipped the back cover of an old Agatha Christie paperback they found on the ground. Today, long after planes have become commonplace, cargo cult beliefs and stories remain embedded in local religious practice.

We have searched for a physical record of an ancient, technologically advanced civilization and found some strange ruins that suggest the existence of powerful engineering abilities that are now lost. But we have not been able to find any trace of a society that could have created the Sphinx ten to twelve thousand years ago, moved the gigantic stones at Baalbek, or constructed Nan Madol out of 458,000 tons of basalt.

The structures are there, but not a trace of writing, not a single tool, not even an apparent reason that something as arcane as Nan Madol would have been built.

So perhaps there is some record of this civilization elsewhere—an oral record along the lines of the stories that were preserved about U.S. Air Force operations in New Guinea by the cargo cultists.

As it happens, such a record may exist. Some of the oldest stories in the world come from the Vedic literature of ancient India and tell of a civilization that possessed scientific skills and technological knowledge, but did not extend to all peoples in the world. There was a war, and it was destroyed so completely that nothing remained.

The central works of Vedic literature are the *Bhagavata Purana,* the *Mahabarata,* and the *Ramayana.* The *Mahabarata* and the *Ramayana* are the oldest, dating from about the sixth century B.C. The *Bhagavata Purana* is about cosmology, and the other two are historical accounts. The texts are thought to incorporate much older material, some of it said to have been recorded as long ago as 3000 B.C. and quite possibly to come from even earlier accounts in the oral tradition.

Throughout this literature, there are accounts of what would appear to be highly technological machines. A careful reading of the texts demonstrates not only that most of the descriptions could fit real technology, but also that the most powerful technologies are described in the oldest literature.

The *vimana* aircraft are the best known of these, and they range from devices that could convey souls through transcendental realms to man-made wooden craft that used wings.

Some of the descriptions seem extremely fanciful, but the more our own technology advances, the more sense they make. An example of this occurs in the Tenth Canto of the *Bhagavata Purana,* where the story of a *vimana* obtained from the God Siva by an earthly king, Salva, is recorded.

This device had some remarkable properties. It could appear as many objects. It could become invisible. Its exact location could not be determined. It never stopped moving, not for a moment.

A modern military aircraft using electronic countermeasures can produce these same effects on an enemy radar screen right now.

Salva also possessed arrows that homed in on sounds. There are missile guidance systems that work by homing in on the distinctive sounds of jet turbines. They have the advantage of being impossible to defeat using normal countermeasures.

Teleportation is also described in the account of King Salva's adventures. Such a thing has until recently been assumed to be impossible. But in a report in *Nature* magazine in 1997 (vol. 390, p. 575) a team of scientists announced that they could cause quantum duplication of a photon, or particle of light, and there is serious speculation that quantum movement of larger objects may be possible.

The projection of illusions that look like real objects is also a feature of the account. Using holographic projectors, defense specialists have been experimenting for some years with creating elaborate illusions of planes against clouds and even vehicles and troops moving across the land.

Should these accounts be dismissed as works of the imagination? Certainly that's what we would normally do, and what seems rational. But it is actually a more serious question than it seems, because of the issue of whether technologically uninformed people could ever have imagined such things at all.

If they could, then, it would seem, we would find such descriptions throughout ancient literature. Strange weapons, flying craft, and imaginative technology ought to abound.

But a survey of the literature of other cultures reveals the opposite. With very few exceptions, ancient literature does not describe machines; it describes powers or beings of power. A famous exception is the account of the fiery wheel that occurs in the Bible, in the Book of Ezekiel. While it's obviously not the only account of a "divine machine," the fact that it is the only one in the Bible attests to the rarity of this subject in ancient literature. Of course, this isn't too surprising, given the fact that ancient people were rarely exposed to machines. The authors of the Bible would never have seen anything fly except birds. They would not have been aware that jet or rocket power can cause something to rise off the ground. This is why the description of the device in Ezekiel may well be the same sort as that of the *vimana* craft of the Vedas—a confused account of a real object of some kind.

Another tradition in which flying machines appear is the fairy cult of northern Europe, but these are never associated with any kind of motive force, but fly by magic.

By contrast, a whole array of different machines are described in the Vedas. In fact, if a primitive people were exposed to Western aircraft, television, electric light, and powerful explosive weapons, they might well write descriptions of them similar to what is found in the Vedas and understandably attribute them to gods who looked human.

If they are not works of the imagination, then the Vedas record a relationship with a technological culture that might have had more than one warring society within it. The relationship the Vedas describe between ancient Indian peoples and this society, or group of societies, was not very different from what many Third World countries experienced during the Cold War, when the Soviet Union and the United States competed for their allegiance by giving them technological booty and constructing magnificent facilities for them in return for strategic alliances.

Vedic lords would worship various demigods and in return get extraordinary craft and weapons from them. They would get help in their wars and would become allied with one side or another. King Salva sought the help of Siva to destroy a city of Krishna's. He was given access to weapons that could affect the weather, and such weapons are not infrequently mentioned in the Vedic texts.

Missiles, rockets, possibly even jet power and bombs, are described in the Vedic literature. The power plants of *vimana* craft emit vapors and create roaring noises. Explosions are described in some of the older texts. But where did this information come from? How was it imagined?

There is no real answer to that question. All of the political alliances and all of the violence that is implied in the Vedic literature suggest that there was more than one advanced society, possibly as many as five or six, and that at least some of them were so advanced that they could leave the earth altogether . . . and that they were not at peace with one another.

In any case, whoever they were—if they even existed—they have certainly departed this world, leaving behind them only a few magnificent structures that silently defy our own science with the riddle of their construction.

Sometimes the most telling facts are small things, and that is true in this case. One could go on and on examining the evidence, trying to get a final answer. But there is one small fact that suggests that it all may be quite true—that there was once magnificent power among men, and that it was all lost in conflict.

When archaeologists dig down to the very bottom of some of the most ancient cities in the world, passing through layer after layer of construction at Ur and Nineveh and other locations in the Middle East, what is found is a material that is also found at ground zero of atomic explosions: fused silica—sand that has been heated to such extreme temperatures that it has turned to glass.

Are there shadows hidden in that ancient material?—memories, perhaps, of a great and terrible age that is gone? If so, the old legends conceal the meaning and definition of our own epoch, a time during which we have climbed painfully back to a certain level, only to find that we are hanging by a thread over the abyss of an equally uncertain future.

But what about the people who had those magnificent powers? What sort of scientific knowledge might they have possessed? If we listen to the music of the old myths, if we let it sing to us and haunt us and capture our hearts, maybe then we will touch the secrets of that science.

9

The Emergency Develops: The United Kingdom

FROM: ALEX RICH, THE HADLEY CENTRE FOR
        CLIMATE PREDICTION AND RESEARCH,
        U.K. METEOROLOGICAL OFFICE
TO: BOB MARTIN, NATIONAL SEVERE STORMS
        LABORATORY, U.S. NATIONAL OCEANIC
        AND ATMOSPHERIC ADMINISTRATION

BOB:

 WE HAVE A SITUATION DEVELOPING HERE
THAT I BELIEVE TO BE UNPRECEDENTED. AS YOU
ARE AWARE, THERE IS A VERY SUBSTANTIAL

FRONTAL SYSTEM MOVING RAPIDLY SOUTHWARD THAT APPEARS SOMEHOW LINKED TO THE TRANS-ARCTIC DISTURBANCE WE HAVE ALL BEEN OBSERVING FOR THE PAST WEEK.

THE STRENGTH OF THIS SYSTEM SHOULD BE APPARENT WHEN I TELL YOU THAT YORKSHIRE EXPERIENCED ITS FIRST TORNADO AN HOUR AGO, AND IT WAS A C-3, NO LESS. THERE WERE DEATHS; WE DO NOT HAVE NUMBERS.

WE ALSO HAVE A DOPPLER OF THE STORM, WHICH IS ATTACHED TO THIS E-LETTER AS A GIF. YOU WILL SEE THE CLASSIC FORMATION, UTTERLY FAMILIAR TO YOU, I HAVE NO DOUBT. BUT NOT SO FAMILIAR TO US HERE. IT'S LIKE SOMETHING OUT OF A TEXTBOOK COME TO EVIL LIFE. THERE ARE VILLAGES SHATTERED, PEOPLE WITH NO IDEA OF WHAT HAPPENED TO THEM. THE MINSTER, WHICH IS OUR GREAT CATHEDRAL IN THE CITY OF YORK, HAS HAD ITS WINDOWS BLOWN OUT. THIS IS NOT A PLACE FOR SENTIMENT, BUT I DO RECALL HOW MUCH YOU WERE TAKEN BY THEM WHEN YOU AND MARTIE CAME. GOD, THAT WAS A GOOD TIME. PLEASE DO GIVE MY DEAR SISTER MY BEST, AND REASSURE HER THAT INTERESTING WEATHER NEVER STRIKES WHERE I MIGHT BE ABLE

TO VIEW IT FIRSTHAND, SO SHE IS NOT TO
WORRY.

MY QUESTION IS, WHAT DO YOU THINK THIS
THING WAS DOING IN YORKSHIRE, AND IN MID-
WINTER NO LESS? WE HAVE NO THEORETICAL BASIS
AND ARE AT A LOSS TO ISSUE A STATEMENT. NATU-
RALLY, THE MET WANTS ONE IMMEDIATELY, SO I
NEED A QUICK RESPONSE IF YOU ARE AT ALL ABLE. I
HAVE ALREADY CONSIDERED THE IDEA OF CALLING
IT A FREAK SYSTEM, BUT I THINK THAT WE ARE IN
MORE TROUBLE THAN THAT.

PLEASE READ ON, BECAUSE I AM GOING TO
TELL YOU QUITE AN EXTRAORDINARY STORY. FIRST,
PULL UP THE LATEST OFF THE MET SITE. YOU CAN
SEE THE COMPLEXITY OF THE SITUATION, WITH
THAT MASS OF EXTREMELY COLD AIR DIPPING
DOWN FROM THE ARCTIC. LOOK AT THE TEMPS.
HOW CAN WE EXPLAIN THIS? THOSE ARE ARCTIC
EXTREME CONDITIONS. ALONG ROADS IN
SCOTLAND, PEOPLE HAVE BEEN FROZEN TO DEATH
IN THEIR CARS. ANYBODY OUTSIDE IS LIABLE TO BE
KILLED IN TEN MINUTES UP THERE. MOUNTAINEERS
ARE MISSING. VILLAGES IN THE NORTH OF
SCOTLAND ARE NOT RESPONDING. THE PROBLEM
MAY BE POWER, BUT I FEAR THE WORST. WE DON'T

HAVE A DEATH TOLL, BUT THERE IS ONE, YOU CAN BE SURE, AND IT IS DOUBTLESS GROWING.

THE LACK OF EMERGENCY EQUIPMENT ABLE TO COPE IS A SERIOUS PROBLEM. WE CANNOT GET AIRCRAFT IN DUE TO THE GALES. I AM NOT UP ON IT, BUT I BELIEVE THAT SOME SORT OF MILITARY ACTION IS PLANNED, AN ATTEMPT TO REACH PLACES THAT HAVE BECOME ISOLATED AND ARE WITHOUT POWER.

IT ISN'T OBVIOUS FROM THE PRESS. I KNOW THAT. DON'T BE DECEIVED. YOU ARE LOOKING AT AN APPALLED SILENCE, REALLY. WE ARE GETTING MINISTERIAL QUERIES BY THE MINUTE, ASKING FOR THIS AND THAT FORECAST. BUT WE CANNOT GIVE THEM FORECASTS. WHAT UNGODLY MONSTROUS FORCE IS DRIVING THAT ARCTIC AIR? I THINK I KNOW, AS I WILL EXPLAIN IN A MOMENT. IF I AM RIGHT, THEN MAY GOD BE WITH US.

LONDON IS STILL RELATIVELY NORMAL, BUT THE CITY IS BRACING FOR WHAT IS EXPECTED TO BE A DROP TO WELL BELOW 0°C. BY MIDNIGHT TONIGHT.

WHAT I DO NOT HAVE IS DATA ON THE GULF STREAM, AND NATURALLY WE ARE ALL WONDERING IF YOU MIGHT NOT HAVE SUCH, OR ACCESS TO IT

PERHAPS THROUGH YOUR NAVY. EVEN IF IT IS CLAS-
SIFIED, AND IT COULD BE IF THEY USE IT IN SONAR
CALIBRATION, PLEASE DO GET IT AND SEND IT OUR
WAY. THE ROYAL NAVY HAS PRETTY EXTENSIVE
DATA, OF COURSE, BUT IT IS MOSTLY FOR THE
NORTH SEA AND THE SHELVES. WE NEED RATE OF
FLOW FOR THE MID-ATLANTIC. WATER
TEMPERATURE. SALINITY. I AM OUT OF MY DEPTH
(SORRY!) WHEN IT COMES TO MATTERS OCEANIC,
BUT IF YOU HAVE ANY MODELS THAT WE MIGHT BE
ABLE TO PLUG IN—WELL, PLEASE GET IN TOUCH
ABOUT THEM ASAP.

IT IS MY PERSONAL BELIEF—AND I WOULD
NEVER EVEN SPEAK THIS OUT LOUD AROUND THESE
PARTS, OLD FRIEND—THAT THIS IS THE BIRTH OF A
TERRIFIC CATASTROPHE THAT WE ARE WITNESSING.
I THINK THAT SOMETHING HAS HAPPENED TO THE
NORTH ATLANTIC CURRENT. I THINK THAT AN
ENTIRE SHIFT OF CLIMATE IS UNDER WAY. THE
NOTORIOUS "CLIMATE FLIP."

IF THIS IS TRUE, THEN ANYTHING, ANYTHING
YOU CAN TELL US IS URGENTLY NEEDED, BECAUSE,
AS YOU KNOW, THE EUROPEAN SITUATION WOULD
BE VERY CONCERNING, ESPECIALLY WITH REGARD
TO US.

I DO NOT BELIEVE, UNDER THESE CONDITIONS, THAT THERE WOULD LIKELY BE A CROP THIS YEAR, NOT IN THIS COUNTRY AND NOT ELSEWHERE IN NORTHERN EUROPE. FOOD SHORTAGES. PERSISTENT COLD INTO THE SPRING. CONSEQUENT FUEL SHORTAGES AND ECONOMIC DISLOCATION.

GIVEN THIS, I SEE A NEED FOR SOME SORT OF EMERGENCY CONFERENCE, PERHAPS UNDER THE AEGIS OF THE WMO—BUT SOMETHING THAT WILL BE ABLE TO ADDRESS THE URGENCY OF THE SITUATION.

LOOK AT THE MAP. HOW EXTRAORDINARY. IT'S AS IF THE ARCTIC IS SUDDENLY EXPANDING, AS IF IT IS SLIDING DOWN THE PLANET INTO EUROPE. WELL, THAT'S A COLLISION WE CANNOT HANDLE.

REVIEW THE DATA THERE AND PLEASE DO RESPOND AT YOUR EARLIEST CONVENIENCE.

BEST, AS EVER,

ALEX

PS: GIVE JENNIFER AND ROBBIE THE BEST WISHES OF THEIR UNCLE, AND TELL ROBBIE THAT THERE ARE SALMON WAITING FOR HIM YET IN THAT STREAM WE VISITED LAST. (I DO HOPE.)

10

Some Answers and
a Huge Question

AS, OVER THE COURSE OF THIS BOOK, WE COME
to understand how close sudden climate change
could be, it will become more and more clear just
how urgent our situation actually is. Before we
turn to the current climatological situation,
though, we would like to explore the message of
the past a little more deeply.

To do that, we are going to turn in an unexpected direction. When we asked ourselves what
the past might have left behind, we answered with
another question: what, from the past, is old
enough to matter and also useful in pinpointing
dates and times?

The answer was obvious: a calendar. But which
one? The ancient world was full of calendars. But if

an attempt was being made to communicate over a long period of time, and to do so to the broadest possible number of people, two things would be necessary. The calendar would have to measure a very long time, and it would have to be universally available.

There is only one calendar that qualifies, which brings us back to the zodiac. It is a long count stellar calendar. To examine its functions as a tool of time measurement, it is not necessary to deal with the issue of whether or not it is prophetic. For our purpose, it will prove to be astonishingly so, but not because of esoteric content. It will prove prophetic because of the way it is designed to mark the ages.

It counts the amount of time it takes for the North Pole to slowly move backward around a complete circle—a total count of twenty-five thousand nine hundred and twenty years. It is divided into twelve segments or signs of twenty-one hundred and sixty years each. Over the course of its full transit, true north will pass slowly through each one of the signs. This effect is caused by the fact that earth's poles are at a slight angle in relation to the sun and is known as the precession of the equinoxes.

The constellations of the zodiac appear to be arbitrarily created, their names different in different cultures, their configurations having only a coincidental connection to the figures they describe. Gemini, for example, is indeed twin stars. But Taurus is no bull and Virgo is hardly somebody pouring out a jug of water. The constellations are a mnemonic device, there to keep us from forgetting the calendar.

Why should people remember it, though?

We believe that we have discovered a possible reason—at least, one worth speculating about. If we're right, then the zodiac reveals something incredible about human affairs and suggests that its creators had enormous insight. If the two of us are not simply looking into the mirror of our own imaginations, there may be some extremely useful information to be gained from this exploration.

It appears that the arrival of each age of the zodiac may have been commemorated by the creation of a great monument, but not always a physical one. Each of these monuments expressed the sign under which it was created and seems to have defined the age in which it was made. Over time, they have become increasingly subtle and spiritual . . . and powerful. Indeed, the most powerful of all the monuments is still richly alive today. In fact, it is the foundation of our civilization.

The first of these monuments that was identified as such was postulated by Graham Hancock, who observed that the constellation of Leo rose directly behind the Sphinx in approximately 10,500 B.C. Could it be that the Sphinx, a representation of a lion with a human head, was constructed to somehow identify the Age of Leo during which it was apparently built?

But why do this? Why decide that a particular sign of the zodiac should identify a particular two-thousand-year period? This returns us to the question, Is astrology a superstition or the remains of an ancient predictive science? If it is the former, then does it measure some sort of subtle influence based on the movements of planets and stars, or does it use their positions as indicators of the progress of a large-scale process of change that is hidden in society itself? If so, this might mean that the zodiac's creators thought that cultural change unfolded according to laws of some kind that could be understood and manipulated.

Maybe, therefore, they understood that our age would parallel this earlier period and thus that we would face the same catastrophe that destroyed them.

Much has been written about the evolution of species. There do appear to be laws governing this process, although their exact functioning remains to be described. Could it not be that these same laws apply to all evolution, including the evolution of culture? If so, then maybe the zodiac isn't magic but science and science that has the potential to be understood. Maybe it contains a message that we can not only comprehend but use, if we can just avoid the traps of either embracing it as magic or dismissing it as bunk.

So let's take a journey across the half of the zodiac that has unfolded since the Age of Leo and see if we can perhaps gain some insight from it.

First, we need to go back to the time when the Sphinx was built, and the place. In those days, Egypt was radically different. When the Sphinx was constructed, there was substantial rainfall in the area. As Leo ended, the weather changed dramatically—it got suddenly colder, and this temperate area became a desert. The whole upheaval took place as Leo transited into Cancer.

After Leo came Cancer, then Gemini, then Taurus and Aries (the precession moves *backward* through the signs of the zodiac). Then Pisces brings us up to modern times. We are exactly half the zodiac away from Leo, or approximately 12,960 years.

For our journey, we will not use modern interpretations of the signs, but rather the oldest ones we have been able to find.

To begin with Leo, there is a very ancient riddle applied to the Sphinx. It asks the question, What has the haunches of the bull, the claws of the lion, and the head of the man? The ancient Greek play *Oedipus,* by Sophocles, contains an early expression of this riddle, although it was written nearly seven thousand years after the Sphinx was carved. The answer, although never given in the play, was "a man, for a man is the measure of all things."

A man, or historical mankind. For it cannot be denied that the foundations of what would become human history were laid down in the Age of Leo. It was during this time, thousands of years before Christ, that human beings in the Western world first began to move beyond the limits of tribal life. Trade began, occasional sea voyages were undertaken, and the earliest beginnings of transmitted memory took place. Also, there may have been a superstorm at the end of this period, and its effects may be recorded in legend. It was probably during this era, for example, that the oldest saga in Western imagination, the *Epic of Gilgamesh,* began to be recited. This epic described the struggle of the hero Gilgamesh to survive a great flood by floating on a raft. It is the prototype of the story of Noah in the Bible.

Do the legends of the expulsion from Eden also date from this period, when the Middle East became a desert? As their temperate world withered beneath the brutal sun, it must have seemed very much to the peoples of that area that they were being driven from a garden by a fiery sword.

Is it possible that man first gained self-knowledge, "the fruit of the forbidden tree," that would lead him into becoming a civilized being at this time? If so, one might ask, then who carved the great sculpture, if man was not yet civilized? Rather than answer that question directly, let's move deeper into the precession of the equinoxes, to the next sign, that of Cancer.

Sir James George Frazer created in *The Golden Bough* an immensely accomplished study of the earliest organized human religions. Drawing on a deep investigation of myth, folklore, traditional stories, ancient literature, and archaeological evidence, he concluded that prehistory was universally characterized by the advent of goddess religions, displacing the shamanic cults of the earliest hunter-gatherers.

Vestiges of the ancient faith lingered on in Greek and Roman religions. Athene was the patron of Athens. She had the traditional three parts of the goddess: wisdom, courage, and strength. Athene's totem animals were different, but they reflected the same basic concept of three-in-one that was memorialized in the Sphinx. In Rome, the goddess Vesta was at the heart of the state religion, and it was believed that if her sacred fire was ever extinguished, Rome would fall. A haunting reflection of the great goddess persists into modern times in the image of the Blessed Virgin Mary, a dynamic and living part of the oldest Christian denomination, Catholicism.

The goddesses of Greece and Rome, and in the present time the Virgin Mary, are very late manifestations of this ancient religion, which Frazer felt must have had its origins prior to the advent of the earliest civilizations, before a clear connection was established between sexual activity and pregnancy. He points out that the earliest religious traditions seem to have been based on the belief that the action of the moon upon the menses was what caused children to be born.

Interestingly enough, this religion would have been ascendant during the Age of Cancer, which is also the sign of the mother goddess. So it was under the sign of the Sphinx that mankind, at least in the Middle East, was forced to end a long period of easy hunting and gathering and begin to gain the rudiments of civilization, and under the next sign, Cancer, that the first great universal religion developed, the worship of the goddess who granted humankind the children needed to continue the race.

The human journey since then can be seen as a long struggle toward an ever more coherent picture of God, as if man had been awakened in Leo and shown his strength, then set on a long journey of discovery.

The next age, Gemini, was when recorded history began. This period, judging from the way religions evolved, was the era during which mankind gradually came to understand that sexuality and pregnancy were connected. Male gods began to emerge and Gemini, the twins, continued for approximately another two thousand years, during which the goddess religions gradually contracted into a few centers, such as Sumeria, where they continued to dominate into the early part of the next age. However, powerful new male gods were ascending in the mind of man, the greatest of them among the Egyptians. Moreover, a new kind of religious practice was emerging. We were not only worshiping our gods, we were telling ourselves moral stories about them and learning to take the lessons into our lives.

Among the earliest of these stories was the tale of Isis and Osiris, which was derived from the ancient idea of the female as the life-giver and res-urrector. Osiris, Isis' brother, was killed by the jeal-ous Set and cut into pieces. Isis was able to recon-struct them and give Osiris renewed life, after a fashion. This idea was preserved down into histori-cal times, reappearing as the powerful story of the death and resurrection of Christ, which has remained central to Christian religious belief into our own era.

The story of Isis and Osiris marked the end of the Age of Gemini, as the resurrected male god of the archaic hunter-gatherers, the people who had walked the earth prior to the coming of the Age of Cancer, reasserted himself. Under the next sign, Taurus, the male gods gained ever more complete ascendancy, but always tempered by the female influence that lingered from the age of goddesses. It was during this era, about 4000 B.C., that the Sumerians first wrote down the Epic of Gilgamesh.

This was also the first era since Leo during which there was the suggestion of a planned rela-tionship between the fundamental ideas of the period and the astrological sign under which it was developing. In this case, it did not take the form of a single monument, but was the introduction of the concept of the bull into evolving religions.

The bull imagery persisted into the first millen-nium of the next age, that of Aries. By that time it was already very ancient. Many of the old god-desses were portrayed with bulls as their consorts, such as the very ancient Sumerian goddess Inanna and the Greek Europa. The wife of the legendary Cretan king Minos, Pasiphaë, fell in love with the bull of the god Poseidon. In his book *Primitive Mythology,* Joseph Campbell says that the image of the bull emerged in the early Bronze Age (during the transit from Taurus to Aries), with its influence extending from the Indus Valley to England.

Many bull statues exist from this period, most of them showing animals with placid expressions, reflecting the fact that the bull gods were thought of as the consorts of goddesses. There is a large statue of Nandi the bull at the Tanjor Temple in India that bears this expression. In that culture, Krishna, the god-man who is central to Hindu mythology, is thought of as the avatar of the Age of Taurus. His mythological helpers are known as *gopis,* or cowherd-girls, in northern India. Cattle persist as sacred beings in modern Hindu belief, a holdover from the days of Taurus.

The next two signs are Aries, the ram, and then Pisces, which are associated with powerful Western religious symbols. The Old Testament contains more references to the ram (55) than to any other animal. The Old Testament was written during the middle centuries of the Age of Aries and is, on one level, the story of the ultimate replacement of the old goddess with a new masculine god.

This process began much earlier with a series of invasions, as northern tribes with exclusively male gods moved south into the Indus Valley and the Mediterranean Basin. They ended the era of the seasonal kings, who ruled for a time but ultimately were sacrificed by female priestly cults. In his book *Occidental Mythology,* Joseph Campbell describes this transition as a major social trauma. Aries literally battered old Taurus to death. Taurean goddesses—for example, Kali, who was associated with bliss, and Medusa, who had previously had a mild disposition—were transformed into monsters.

At the same time, civilization reached ever higher levels of organization. At the beginning of the Age of Aries, the first great cities of our historical epoch appeared in Sumeria, Egypt, and the Indus Valley. Toward its close, the Jews colonized the land of Canaan and became the Israelites, and the idea of the single god finally found its place in human culture. The Jews began to worship the most sophisticated god yet conceived: a formless, eternal, and yet deeply personal essence called Yahweh.

Out of this worship came a whole new social order. Moses gave the Jews a set of commandments that turned natural morality into written laws. The stage for the future was thus set during the Age of the Ram. It was the emergence of the Jewish god that would provide the symbolic focus of the next era, the Age of Pisces, at the end of which we now live.

Pisces had a living avatar in Jesus Christ. He appeared shortly before the actual turn of the calendar, during the period when Aries was winding down. The deep connection between Christ and the zodiac is revealed by the fact that his earliest symbol is the fish, he is called "the fisher of men," and his ministry began early in the Age of Pisces. His apostles are drawn from a group of fishermen, and early Christians identified themselves by the symbol of Pisces the fish, a practice that Christian fundamentalists have revived, but without realizing the astrological origin of the symbol.

Christianity mixed the humane tenets of Greek thought with the moral rigor of Judaism, emerging as the most ethically directed and yet compassionate religion yet seen.

From the Sphinx to Christ, then, there have been a long series of enormously important cultural events and symbols that have been associated with the astrological signs that they have happened under—unless we are reading the past too carefully, which could always be the case. We find these speculations fascinating and possibly useful, but as the kind of rigorous analysis needed to establish their objective value has not yet been done by cultural anthropologists and other scientists, we cannot be certain of our ground. Nevertheless, the speculations seem potent to us.

Each sign has offered deep, even defining meaning to its age. By reading them in the context of only the simplest and most traditional interpretations of their meaning, what appears to be a level of planning in human civilization that is literally deeper than history is revealed. And the reason that a calendar as long-term as the zodiac would have been invented becomes clear: it is a template being used by whoever it is that is influencing human life on this massive scale.

So what does it tell us, as we leave the Age of Pisces and enter the Age of Aquarius? Christian civilization is the fish that has been swimming happily in the water of faith.

Until now. Because now everything is once again changing. Yet again, new beliefs are challenging the old. The eighteenth century saw the beginning of our transition to the next age when rationalism began to challenge faith. In the nineteenth century, Christianity began to contract, as more and more people started to see the world around them in terms of reason and science.

Now there are major environmental disruptions all around us, and as our situation gets more desperate, it becomes less and less clear that prayer is the best tool to use against the relentless mathematics of something like global warming. Can a world that just experienced the senseless murder of two hundred million human beings in a thirty-year upheaval that involved two world wars really count on or believe in a personal god who protects us and is concerned for our welfare?

Those of faith say yes, but the average person looks at the dangers that surround him and wants something more practical and assured: effective science, dynamic government, a social order willing to face problems and work toward their solution.

It is getting hotter and hotter, the weather is getting stranger and stranger, and the future is looking dark. All around us, huge numbers of species are going extinct, diseases are on the rise, the ozone layer is thinning, every solar maximum (peak of sunspot activity) seems more powerful than the last, and the environment is becoming more and more hostile to life.

In other words, the action that characterizes the beginning of Aquarius has started: the pond of belief and assurance in which Pisces swam so comfortably is being drained, its living water being poured out by Aquarius. We who have always relied on earth's ecosystem to sustain us must now find a way to sustain ourselves.

But how can we? We're not made to breathe this harsh new air. The Age of Aquarius will not be the new age of directionless freedom that has been foretold in popular culture, but rather a period of intense search, as mankind seeks to somehow do what ancient fish did back at the beginning of time, learn to survive outside of the ocean.

The Age of Leo is precisely halfway across the zodiac from the Age of Aquarius. It was during Leo that the last great catastrophe took place, which led to the floods and mayhem recorded in our earliest history, and also to the rise of human civilization, which began when we were forced to replace lost natural food sources with agriculture.

Thousands of myths and legends from around the world tell of an environment that stays stable for long ages, but then suddenly falls apart, causing destruction on such a gigantic scale that entire ages of man are obliterated.

Let's see if we have, perhaps, decoded the secret of the zodiac and read the message that our forebears left for us. We must ask the question, Where do we stand right now?

We have already seen that something touched off a massive change in the weather back in prehistory. Flood myths from all over the world tell of a period of devastation, and in later chapters we will show just what sort of catastrophe those myths may be describing.

We can't go back to the critical years before the catastrophe unfolded, however, and observe the month-to-month details of climate change. There simply isn't enough data. What we can know is that the shift was extremely sudden. But what led up to it? Would there not have been dozens of small but significant changes before the catastrophe actually occurred? The fossil record tells us that climatic changes are like that: they seem to build over many years, ending in a sudden explosion.

Our ancestors warned us about Hamlet's mill, and their warning turns out to reflect with uncanny accuracy scientific knowledge about the way climate changes. Hamlet's mill turns for long periods with great smoothness. Then, suddenly, an explosion.

The great majority of our century's weather catastrophes have happened within the past thirty years—a blink in geologic time. The most deadly cyclone of the century hit Bangladesh in 1970. Perhaps a million people were killed. The heaviest hailstones ever recorded, upwards of two pounds each, hit that same country in 1986. Recent hurricanes such as Mitch and Gabriel, which had a storm system *above* it that reached high into the stratosphere, are indications that catastrophic weather systems are getting more powerful. Floyd, in 1999, was at one point the size of Texas.

The long *El Niño* of 1997–98 has given way to abrupt cooling of the central Pacific, *La Niña*, and there are indications that another *El Niño* event will start soon. This means that the Pacific is now wobbling between temperature extremes, another symptom of growing instability.

The year of 1998 was one of absolutely extraordinary weather upheaval. Droughts caused massive fires in Mexico, Brazil, Indonesia, Florida, southern Europe, Australia, and Siberia. New Guinea had its worst recorded drought. The Panama Canal became so low due to drought that it was too shallow for many ships. East Africa endured its worst floods in forty years. Tibet had its worst snows in fifty years. Ice storms devastated trees from Maine to Quebec. Cocoa and rubber crops failed in Malaysia; coffee, in Indonesia and Ethiopia; rice, in Thailand; cotton, in Uganda; and fishing, off Peru. The ozone hole over the Antarctic became larger than ever, and the ozone layer over the Arctic thinner than ever. And it was the hottest year on record.

So, where do we go from here? Where is this all leading? This gets us all the way back from the mysterious human past to the mysterious human future. In the past, something devastated the world, especially the Northern Hemisphere. The witnesses of it left myths of flood and mayhem. It probably destroyed a great civilization, or protociv-ilization, that left the world littered with enigmatic monuments.

We will show that Hamlet's mill went berserk in the past. We will show that we are leading up to another of these rare environmental events. Unless something changes—and change is still possible—there is every likelihood that, halfway around the zodiac from its last appearance, the superstorm will come again.

11

The Superstorm Returns

ON ONE SIDE OF THE SUPERSTORM—THE CLIMATE is temperate. After six weeks of destruction that would make a nuclear war look tame, a new ice age may well have begun.

Is it possible? Yes, and as soon as a mechanism is triggered that can deliver enough energy into the atmosphere to generate it, it's going to happen. By definition, a superstorm would involve an entire hemisphere. Its winds would reach extreme velocities, possibly in excess of two hundred miles per hour.

The storm would be triggered by a sudden increase in Arctic temperatures at the surface—exactly the kind of warm snap that could occur at any time during the global warming scenario presently unfolding—combined with extreme cold aloft. The warm flow of air would heat an ocean surface already affected by a loss of salinity due to polar melt and runoff from Greenland. The lack of salt in the water would cause it to take on heat quickly. At that point, the flow of the North Atlantic Current would suddenly change, dropping south.

When this happened, the ultracold air trapped above the Arctic by the warm airflow would slide southward, with a violent outcome.

The storm would last until the ocean cooled enough for the flow of the current to be reestablished. Before that happened, there would be a massive blizzard or series of blizzards that would dump billions of tons of snow across a fifth of the earth's surface. When the sun finally did return, the huge increase in the earth's albedo, or reflectivity, caused by the snow, would cause a dramatic drop in temperature. Whether the ice would melt or persist across the next summer would depend on its depth. If it persisted, a cooling trend of some duration would result. There would even be a possibility that a new ice age would begin.

It isn't unusual for the Arctic to experience a period of slightly warmer weather in late January and early February. But it's not as if you'd notice: temperatures do not rise to the point that you can comfortably go outside without heavy clothing.

This time, they would.

Normally, the enormous amount of energy it takes to drive great storms means that they don't last very long. Hurricanes, for example, decline quickly when they make landfall and no longer have abundant water vapor rising off warm tropical oceans to sustain them.

Blizzards form when warm air moving north collides with a cold air mass that has been pushed down from the Arctic. The rain dropping down through the cold air mass becomes snow or sleet and can be accompanied by high winds, but gusts of more than a hundred miles an hour are rare. This is because there is intrinsically less energy involved in such a storm than there is in a hurricane. As soon as the warm air entering from the south reduces in intensity, the storm dissipates.

In addition, a blizzard is normally more short-lived than a hurricane, which can continue for many days, even sometimes for weeks, as long as it remains over water and ocean conditions can support it. Because the blizzard normally does not develop as distinct a cyclonic structure as the hurricane and depends on a type of airflow that is usually suppressed by stratospheric winds during the winter months, it generally lasts, at most, a few days.

The snow cover such storms leave behind, although it may become tightly packed, is almost always gone by late May from all but the northernmost reaches of the Arctic. Absent this snow to cool the surface and reflect new heat back into space, things warm up during the summer and the world climate we are familiar with continues to prevail.

Nevertheless, the evidence that long-term changes in climate do take place is irrefutable. The ice keeps coming back, and we aren't sure why. But *something* acts as the trigger, and we know that this event is a sudden one.

There has been a certain amount of scientific speculation that ice ages are linked to the amount of sunlight the Northern Hemisphere receives during the winter. Because the earth is on a slightly variable orbit and also "wobbles" slightly because its axis is not precisely vertical to the sun, this amount varies over the years. And there is some evidence that ice ages occur only when the earth is farther from the sun during the northern winter. However, if this were the only cause, ice ages would have been a feature of earth's life forever, and that has not been the case. Ice conditions are rare, in terms of geologic time.

Right now, the earth is relatively closer to the sun during the northern winter, which mitigates against a sudden change to glacial weather conditions. It would seem that the return of the glaciers is the last thing we need to worry about.

But is that really true? To make a great storm that would leave a sheet of ice across a vast area, there must be a source of atmospheric moisture, which does not exist—at least, not at the moment. An absolutely massive amount of water vapor would have to be pumped into the atmosphere of the far north in order to get a storm like this going. In addition, there needs to be a reservoir of air in the stratosphere that becomes much, much colder than normal.

Is there any mechanism that could cause this in the future or that has done so in the past?

In fact, as we will show, we are on the razor's edge of a situation in which enough energy *will* be present to generate this massive storm. If it happens, it will quite possibly be the second superstorm in the past ten thousand years.

So let's see, in the context of the kind of measured scientific speculation that an informed layman can do, whether the conditions that would generate a superstorm could really happen.

For the storm to get going, we need certain things, chief among them lots and lots of water vapor pouring into the atmosphere. Then we need something to fill that water vapor with heat energy, and the impact of a great deal of cold air to release that energy in the form of a great blizzard. We also need, for the storm to get as big as we expect, a factor that will so amplify its circulation that it becomes, at least for a period of weeks, self-perpetuating. This means that the storm must be able to develop a circulation so large that it is directly fed by arctic cold and tropic heat—extraordinary arctic cold and intense tropic heat.

Now that we have seen in the human and fossil records a certain amount of evidence of a violent event occurring at about the time the present warming trend was mysteriously interrupted, we need to find out what we can about the archaic climate.

Do the sediments and ice cores taken from that period tell a story that suggests a superstorm?

To find out, we're going to travel back across the recent geologic past, to see if we can identify the reality behind the floods and catastrophes of legend and the recorded extinctions. Then we're going to take a look at what is happening right now, to see just how similar our present climate is to that period.

But first, we perhaps ought to return to the superstorm that has been developing over the pages of this book and find out what will happen as it reaches its full force.

12

Canada: A Cry for Help

The northern reaches of the world were in danger, they were in terrible danger, and nobody understood it. The world meteorological community was confused. Weather, in its essence, is about air and what happens as it absorbs and releases energy.

Science has identified certain patterns that repeat again and again during this process. We call them by many names— supercells, typhoons, hurricanes, tornadoes, cold and warm fronts, blizzards.

Blizzards are a type of storm that occurs when warm and cold air masses intermix and attempt to equalize their temperatures. Water vapor becomes rain, which turns to snow, which falls gently or hard, leaving the world magic with snow or perishing beneath an agonizing weight of ice.

This time, the storms were not like ordinary blizzards. And every meteorologist who stared at a monitor, watching them boil up into the stratosphere, knew that something was going profoundly wrong above the Arctic Circle. First, the cloud tops were rising far higher than is normally the case. They were reaching well into the stratosphere, developing fantastically powerful circulatory patterns as high-altitude air, made ultracold by the fact that greenhouse gases were now holding so much heat close to the ground, went speeding to the surface. NASA's new satellite-borne wind measurement system could scarcely be believed: surface gusts in excess of two hundred miles an hour were apparently present in some of these cells.

The storms were short-lived, speeding across the flat land and the sea so fast that their own speed tore them apart.

The world meteorological community struggled to understand. But more than satellite data was needed, and surface data from these sparsely populated, harsh regions was hard to come by.

However, the most important information was coming from thousands of miles farther south, where oceanographers were testing a terrifying theory. NOAA had sent a small ship, the fifteen-thousand-ton Ocean Tester II, *into the violent waters off the Grand Banks in a desperate attempt to quickly gain information scientists had been trying to find money to obtain for years.*

There should have been an array of sounders and buoys in the area, but instead there was only buoy 44011—far out of ideal position on Georges Bank—which had sounded the original, ignored, alarm.

The Ocean Tester *had broken up in a brutal squall while dropping scientific instruments called inverted echo sounders*

into the ocean off the southern tail of the
Grand Banks. Eleven men had died, but the
fourteen echo sounders they had succeeded
in deploying were providing an effective
stream of data. That data was causing a
storm surge of its own.

There should have been an easily
detectable eddy field where the Gulf Stream
formed the North Atlantic Current in this
area. But it was not there. The sounders,
which measure the time it takes for a sonic
pulse to travel from the sea floor to the surface and back, were reading low and
diminishing flow in an area where the current should have been strong.

To the oceanographic community, the
conclusion was inescapable: something had
happened to the crucial North Atlantic
Current, something without precedent in
recorded history.

Events in the Southern Hemisphere
suggested that this was part of a fundamental change in transoceanic current flows.
The storms that were battering Australia
and New Zealand meant only one thing: the

change in the currents was profound. The whole planet was affected.

A second ferocious typhoon was already forming in the central Pacific, again heading toward Japan, which was still reeling from the effects of Max.

Specialists were called on from many different areas and disciplines to participate in a colloquy via Internet relay chat and teleconference.

By this time, everybody was fairly sure what was happening. The Atlantic Circulation transports warm, salty surface water to the north and sends cold, dense water flowing south. It supplies deep water through the North and South Atlantic into the Indian, Pacific, and Antarctic Oceans. The effect of all this water transport is to keep ocean temperatures, and thus also air temperatures, in the zone that we're familiar with.

Cold, deep flow into the Pacific had stopped, causing a rapid warming of surface water, suddenly deprived of its heat exchange from below. Thus the monster typhoons.

Meanwhile, warm water was no longer being pumped into the Arctic . . . and that was why the very life of Canada now hung by a thread.

It had been known that the North Atlantic Current could destabilize. Ice cores had proved that it had happened before, even as recently as eight thousand years ago. There had been violent Arctic extinctions then, followed by two hundred years of cold—the result of a superstorm, perhaps, with the cold being caused by the reflectivity of the ice cover it had left behind . . . the melting.

But nobody had thought that it would happen again soon. The massive amounts of energy that would be needed to change that flow made it seem like a problem for the future, not now. There was sound scientific thinking behind this: the Princeton Ocean Model suggested that a critical part of the current known as the meridional overturning circulation was much too strong to be changed except by a much greater alteration in ocean temperature than was predicted by the various global warming models.

But the models had not been designed to reflect what would happen if the polar cap began to melt. They did not reflect the effect of floods of freshwater gushing southward from the North Pole or down the streams and rivers that were roaring with ever greater force off the dying Greenland glaciers.

Better, more exact models of current behavior were under construction, but there were limits in terms of available data and computing power.

Obviously, a dramatic shift had taken place, one that rendered moot all the existing models. Complex modern organizations like NOAA and its many allied services from outside the U.S. do not respond quickly. They're vulnerable to the unexpected. They can be overtaken by events.

But they tried. They did all they could in this incredible situation. Unfortunately, there was a failure of imagination. An ad hoc committee of climatologists decided that the disruptions would play out over a period of four or five years, during which time the public would have time to adjust to new conditions.

It wasn't that they were wrong. The weather would certainly remain in chaos for many years. But they just had not been able to imagine the ultraviolence with which the process would start. It was too far outside of their experience, and all human experience.

Nobody in recorded history had seen nature at its most violent. Nobody could conceive of what this actually meant.

But then something happened that brought every scientist assigned to the problem to a dead halt. People were paralyzed with amazement as soon as they understood the data. There had been a report from a small Inuit village in northern Canada that the temperature had dropped in a matter of an hour by over a hundred degrees Fahrenheit.

This could mean only one thing. But the idea that storm circulation could be so violent that it would draw supercooled air all the way to the surface was a subject for classroom speculation. It wasn't a real-world problem.

But it was. Now.

Scientists at various NOAA facilities in the northern part of the U.S. began quietly arranging for their families to move south, seeking shelter with relatives in Texas and the desert southwest, in Florida and southern California.

Thus began the greatest migration in human history, with a few extra cars joining the normal traffic flow, a few hurrying droplets in the ocean of the ordinary.

Wind speeds were picking up all across the Arctic as one ultraviolent storm after another appeared, their cloud tops bulging to ever more unprecedented heights, their surface winds rising higher and higher.

The U.S. National Weather Service and the Canadian Weather Service both began activating their emergency programs. In the U.S., Emergency Managers Weather Information Network (EMWIN) officials and the Federal Emergency Management Agency (FEMA) began a series of meetings designed to prepare the country for what was now certain to be a massive disruption.

A presidential order was sent to the Immigration and Naturalization Service, distribution restricted, informing it that the U.S.-Canadian border was to be opened for migrants. Emergency shelters and food banks began to be organized. FEMA representatives surprised many a county and school-district official across the country with a call asking that they designate their large-space buildings as emergency shelters. Food and medical supplies began to be moved into place. More discreetly, body bags began to be distributed—but only a few hundred thousand.

Still, nobody really understood. Not even yet.

These preparations were considered foolish by some elements within the government, and it was not long before they became known to the directors of the Blue Foundation, a think tank that was hugely influential in Congress.

Congressmen began asking questions: What was going on, why was all this money being spent, what budgets were involved? The

General Accounting Office was ordered to audit FEMA's storm preparation activities. Fearful of similar treatment, NOAA put EMWIN's plans on hold.

So the food supplies were not delivered to the designated shelters. Hospitals were not informed of the possible crisis. The International Red Cross and the World Health Organization were left ignorant of the fact that the greatest climate change in history was building, and building fast.

The weather, however, was not affected by the actions of the political strategists in the Blue Foundation. The weather didn't care.

The first general call for help came from Canada. Normally, northern Canada expects harsh winters. One Arctic front after another sweeps across the region from October through April. It's cold country, and it isn't going to be surprised by violent weather in February.

But nobody had any experience with what began to happen in the Northwest Territories, among the small villages and towns that extracted their meager living *from a world that had almost nothing to give.*

In recent years, temperatures in the region had warmed so much that the permafrost was melting. As elsewhere in the populated Arctic, buildings were rocking on their foundations and trees were being drowned by summertime root flooding. Last August, temperatures in the region had risen into the nineties for the fourth consecutive year. People fell sick of respiratory illnesses. Molds and pollens choked the air along with vast clouds of mosquitoes. Autumn didn't come until late October. So far, this winter had been a pretty tolerable affair. Temperatures only dropped down to around zero Fahrenheit, and blizzards had been few and far between.

Auyuittuq *means "the land that never melts," but recently not only in the national park of that name, but throughout Baffin Island, that had not been true. In fact, it hadn't been true anywhere in the Arctic. As scientists had predicted in 1998, blue water had opened up over the North Pole last year.*

Now Baffin Island was being wracked by a storm that had blown up so suddenly that the few thousand inhabitants of the world's third largest island were taken by surprise. It came up with awful suddenness, darkening the bright winter sky. The waters of Baffin Bay became so violent that the sea seemed to transform itself into spray and rise into the air. Across the wide island, from Nanasivik to Iqualuit, the winds gusted in excess of two hundred miles an hour. Vast quantities of snow poured down, suffocating animals where they stood, capturing practiced outdoorspeople along streets they knew as well as their own skin, crushing those buildings that had not been shattered by the wind.

Temperatures dropped so low that uncovered skin froze instantaneously. To take a breath of air too fast was to risk death by frost in the lungs. Caribou and grizzlies began to die, and then human beings. Large creatures, unable to shelter in cracks and crevasses, were vulnerable.

In weather centers farther south, meteorol-

*ogists watched as the storm grew, enveloping
the island, then rolling south across the
Arctic Circle like a monster unleashed.
Emergency bulletins began to be issued by
the Canadian Weather Service, but they
could scarcely reach areas whose communications facilities had been destroyed.*

*People whose ancestors had lived in this
hard country for ten thousand years began
to die in numbers. They died on foot,
swathed in warm clothes. They died in
trucks and four-wheel-drive vehicles that
struggled south until the snow stopped them.
A few even died in airplanes, which fluttered
like butterflies in a summer breeze as the
storm overtook them.*

*All died, in the end, as the environment
crossed a line beyond which nothing but
simple creatures like lichens could survive.*

*It was like night in the depth of the
storm, howling, roaring night.*

*And then the series of supercells that
made up the storm organized themselves further. To some of the meteorologists glued to
their flickering screens, it looked as if the*

storms were living creatures, as if they were melding themselves into a greater being, something born of terror. They called the storm Legion, then, for it was many, many made into one.

Canadian authorities saw that it was going to come south. It would have to come south very quickly, perhaps in a matter of days. From Vancouver to Calgary, from Winnipeg to Toronto, the warning was passed: prepare shelter, gather food, organize the population.

But there was only one thing to do, really, something so radical that most people simply could not imagine it: To live, you had to move. You had to go south, and you had to do it immediately. Otherwise, you would die. Such are the awful mathematics of climate and nature's hard law.

13

Critical Cycle

~~~~~~~~~~~~~~~~~~~~~~~~~~~~~~~

NOBODY LIKES TO THINK ABOUT ENVIRONMEN-
tal problems. In part, this would seem to be because we blame ourselves for the condition of our
world. But should we? It seems to us not—at least,
not entirely. While we are definitely affecting the
speed at which things are unfolding, nothing
mankind has done—harmful to the environment
or helpful to it—has changed the fundamental
cycle of destruction that grips planet earth.

Later, as we discuss in more detail how this cycle
works, we will also suggest a massive engineering
project that would give us control over it. This project will be so gigantic that it will seem impossible.
But without quite realizing it, our civilization has
graduated into an era of supermassive engineering

capability. Not only can we plan projects on a transcontinental scale, we have the capability to execute them.

As an example, the engineering ability is there to build a water tunnel from the mouth of the Amazon River in South America all the way across the Atlantic to Morocco, delivering freshwater to the desert. But what happens when the absence of Amazon water spilling into the Atlantic makes the salinity of the ocean increase? And how does the blooming of the North African desert change weather patterns worldwide?

We lack sufficiently advanced environmental models to answer such questions with certainty, but we need them. Such a water tunnel would revolutionize life in North Africa. But would it also destroy the world's delicate environmental balance? Since we cannot be sure, we cannot construct it.

Another example: We could build mirrors in space that would reduce darkness. This was attempted in 1999 by the Russians. Fortunately, it failed.

But it wouldn't even be a particularly expensive undertaking. Street lighting, with all its expense and power usage, would become unnecessary. But would we want to do this? Would we want to give up the night? What about the stars? If we blind ourselves to all but their memory, might not future generations forget that they are even there?

So there is little serious interest in bringing the Amazon to Africa or flooding the night with light. Our point is that we are capable of engineering on this scale, and there is a possibility that we could reengineer our planet to end the persistent climatic upheavals that have probably been our genesis, now that their continuation has become such a danger to us.

However, it is also true that the alteration to the planet that we are discussing would cause enormous environmental changes, and they must be understood first. As the planet is now configured, the gigantic heat-exchange system that is in place acts as a preventive to runaway global warming.

On the surface, the environmental situation we are in now appears worrisome but not immediately dangerous. As stated in *Life* magazine's August 1998 issue, "the consensus among climatologists is that global warming will continue unless we dramatically curb industrial emissions, particularly carbon dioxide, that trap heat which would otherwise escape into space."

The magazine then quotes Jerry Mahlman of the National Oceanic and Atmospheric Administration, who says that we "have about twenty-five years" to start acting to reduce carbon dioxide emissions.

Is the situation really that stable?

The cycle that is responsible for our present climate is an established scientific reality. There is not only strong evidence in the geologic record of what happened the last time, there is also a compelling human memory that has been embedded in myth and legend. This could be one of those times when Hamlet's mill goes berserk.

If this is true, then be warned: We are about to live through a prime moment in the destiny of earth and man. All that is needed to see this is to expand the scale of our normal worldview. There is no need to embrace any speculative science. It is, rather, a matter of learning to think in geologic time.

From a long perspective like this, it looks as if we are involved in the terminal climax of a massive extinction event, one that is unfolding on the truly enormous scale of such legendary extinctions as the Permian event, which destroyed almost all living things, and the Cretaceous, which extinguished the dinosaurs.

To see it for what it is, it's important to understand that although these past extinctions ended suddenly, they were preceded by millions of years of decline.

The event we are living through did not begin a hundred years ago or a thousand or even ten thousand. It began nearly three million years ago and is following a pattern that has been established for most of the history of life on earth.

The catastrophe that almost sterilized the planet at the end of the Permian period, 270 million years ago, began about 2 million years before it reached its climax. Similarly, the impact that ended the age of dinosaurs came only after more than 2 million years of persistent decline.

In this scale, the past fifteen thousand years can be seen as the terminal phase of the present extinction event, and the hundred-year period since the advent of the industrial revolution is eerily similar in its effect on earth's species to the massive fires and pollution that followed the asteroid impact that brought the dinosaurs to an end. Even the rate at which human activity is destroying other species is similar to the speed at which the burning forests and lack of sunlight of sixty-five million years ago destroyed species at that time.

Whether humankind will be among the victims of its own destructive presence on earth we don't yet know. Certainly, we must all do whatever we can to prevent that, and prevention begins with understanding—not only of the true nature of what is happening, but also of what our role might be in changing the situation for the better.

Our journey to this conclusion began in the summer of 1998, during an unusual climatic disturbance. Like everybody else, we had assumed that man was the only significant contributor to climatic shift. Our industries were spewing out too much pollution, with the result that global warming and the introduction of chemical wastes into various food chains were causing die-offs and extinctions.

The signs were everywhere. A powerful El Niño was warming the waters of the mid-Pacific and causing droughts in tropical areas from Southeast Asia to Brazil. The drought band stretched as far south as northern Argentina and as far north as Florida and the U.S. Southwest. Although scientists have identified El Niño as a periodic event that takes place when trade winds change their direction and move warm water from the western Pacific eastward, the increasing frequency and aggressiveness of the events could not be explained. In September of 1998, some scientists were speculating that increasing volcanism beneath the Pacific might be a contributory factor, while others pointed to a more conventional global warming scenario.

Nonetheless, the effects of this El Niño were far greater than any in living memory. There were some fearsome storms connected with it, including a typhoon that struck Guam in June carrying wind gusts of 230 miles an hour, the highest ever recorded. But these upheavals were dwarfed by another phenomenon: terrible fires breaking out in parched tropical rain forests, a phenomenon predicted in *Nature's End* back in 1985.

By June, there was smoke in the atmosphere in Texas from gigantic fires burning a thousand miles away in southern Mexico and Guatemala. Conditions were oppressive: iron gray skies, a sharp scent, red sun at midday, many people suffering from respiratory illness. The smoke cloud was one of the largest in human history, stretching from northern South America all the way to

Illinois. It was exceeded only by the massive cloud that had blanketed Southeast Asia when fires ran out of control in Indonesia. The smoke there had been so thick that there was darkness at noon in Kuala Lumpur, twelve hundred miles away across the South China Sea.

Abruptly, in the middle of the summer, the El Niño ended, to be replaced by its opposite: La Niña. In this situation, cold water replaces the warm water that has been blanketing the central Pacific. Among the many climatic outcomes is a proliferation of Atlantic hurricanes, and the 1998 season was one of the worst in years, ending with the unprecedented devastation of Honduras by Hurricane Mitch.

Given that La Niña followed El Niño so closely, by October of 1998, weather patterns over the Pacific had not been normal for over two years. Also in the autumn of 1998, it was beginning to seem as if another El Niño would replace La Niña in the not-too-distant future, meaning that there might be no normal period at all.

The spring of 1999 brought more wild weather, with tornadoes appearing in such diverse places as China and the United Kingdom. A tornado swarm that struck the United States in June contained one funnel with winds in excess of 313 miles an hour, once again the highest ever recorded.

The past ten years have revealed a new and ominous pattern, which has, since 1995, grown rapidly more intense. In fact, the changes are so dramatic that the presence of powerful voices still claiming that there is "no such thing" as global warming shows that human beings remain as capable as ever of self-delusion.

Signs of rapid mass extinction are growing, and we have to be aware of what really causes it and what it means. It is not impossible that the human food chain will be broken by some crucial extinction. Such breaks can be indirect. An example would be migratory bird populations in North America, which are under increasingly severe pressure from a number of different factors, ranging from lighted cities in their flight paths to weather changes to pesticides in their food. If these birds go extinct, the huge weight of insects that they eat will remain unchallenged at a time when global warming will be causing insect populations to explode. The resultant increase in the number of bugs would be uncontrollable, not without the use of chemicals so dangerous that they would render the food they protected inedible.

Many subtle and dangerous challenges to species survival have begun to develop in the latter half of our century. A fatal combination of negative environmental changes has appeared and is quickly growing more serious.

In the eighties, the layer of protective ozone high in the atmosphere above Antarctica began to thin, then to show holes. Without it, ultraviolet light reaching the earth's surface increases. Reduced ozone seriously disturbs the growth of plants, especially food species that have been bred to grow fast. Exposed animals are at risk for eye damage, skin cancer, reduced immune response and genetic problems.

In the late eighties, the whole planet's ozone layer was observed to be thinning. Holes appeared over the Arctic. Thinning became severe over Europe.

The process was thought to be primarily due to chlorofluorocarbons, a type of chemical widely used in the manufacture of coolants. International efforts were made to reduce chlorofluorocarbon emissions, with the result that they were in measurable decline by 1995. However, in September of 1998, it was announced by British and Australian researchers that another ozone-destroying chemical, Halon-1202, was rapidly increasing in the upper atmosphere, and ozone destruction was continuing. At the same time, it was revealed that the source of this chemical was not known. It is possible that it is a by-product of the production of Halon fire extinguishers in China or perhaps military activities around the world, but nobody is certain.

In 1998, levels of protective ozone in the upper atmosphere became alarmingly depleted. For ten years, Australia had been issuing warnings about the danger to people of overexposure to the sun, and this procedure was beginning to spread around the world.

Between 1975 and 1991, the incidence of skin cancer in men, except melanoma, increased 812 percent, according to the U.S. National Cancer Institute. During the same period, melanoma increased 66 percent, and the mortality rate from this disease went up 30 percent. The rates were much lower in women, which was attributed to their correspondingly lower exposure to direct sunlight due to the fact that fewer of them work outdoors.

But for a few studies with uncertain conclusions, whether the increased UV reaching the earth was having a destructive effect on animals and plants remained unclear. However, a dramatic increase in sickness among wild species and herd animals in various parts of the world has been observed. From widespread sicknesses among frogs and other amphibians to mad cow disease in England to a rabies pandemic among the forest creatures of the eastern United States, animal sickness seems to be on the rise everywhere. Sheep living in southern South America, where long periods of very low to no ozone cover were taking place, are suffering eye damage.

In part because exposure to ultraviolet light reduces the ability of animals to ward off disease and in part because of changing climates, new diseases began to appear and old ones became more dangerous.

In April of 1998, it was reported that an unusually powerful bacterial plague was killing coral in the now chronically overheated Caribbean. So-called "sick pond syndrome" was decimating populations of frogs, toads, and salamanders world-wide. Half a dozen different diseases are attacking these species, apparently because their immune systems are being stressed by a combination of pollution and excessive heat. The extent to which pollutants are entering the bodies of animals was revealed by a discovery of Dutch scientists reported in August of 1998. They found that polybrominates used in flame-retardant fabrics were appearing in the livers of whales that feed only in the deep ocean. These chemicals behave much like DDT when they are eaten.

Not all species were being threatened with extinction. In fact, some were thriving. Combined with the underlying warming trend, the El Niño made 1998 the warmest year ever recorded. As a result of this, mosquito populations all over the world exploded. There were what the World Health Organization described as "quantitative leaps" in malaria cases around the world. Thousands were infected with Rift Valley fever in Kenya, resulting in two hundred deaths, after the heaviest rains since 1961—attributed to El Niño—fell in the area. The incidence of cholera rose in Latin America and parts of Africa, and in the U.S., a proliferation of deer mice due to abundant desert rains placed large numbers of people in the Southwest at risk for Hanta virus. In August, Houston, Texas, began combating swarms of mosquitoes by insecticide spraying, and New Orleans was enduring a plague of roaches. Wetter weather, attributed to climate changes, was blamed in both cases. After an unusually warm eighteen-month period, drug-resistant tuberculosis was found to be spreading in Russia, under conditions of social breakdown that make containment of the bacterium unlikely. In 1999, New York City began to spray for mosquitos after an outbreak of encephalitis, and Laredo, Texas, was contending with mosquito-borne Dengue fever. On Long Island, cases of malaria spread by local mosquitos appeared.

Extinctions of local animal populations were under way worldwide. *The Global Biodiversity Assessment* presented by the United Nations Environment Program in Jakarta revealed that the rate of extinction among flowering plants and vertebrate animals was fifty- to one-hundred-times greater than expected just a few years prior to the creation of the report.

It seemed that the stage was being set for another mass extinction event—in fact, the climax of the one that has been going on since long before mankind even appeared.

14

Beyond Gale Force 10

*To understand the scenario as it is now
unfolding, we need to set the climatic stage.
The year of the superstorm, a summer
weather pattern will persist over the
Northern Hemisphere well into autumn—as
it often does right now. The polar cap and
the Arctic glaciers will continue their melt
into October.*

*As the surface warms, extreme cold builds
in the stratosphere. This is because greenhouse
gases accumulating near the surface trap*

more and more heat, preventing it from reaching the high atmosphere. This problem was first noticed in 1998 and has continued to get worse every year. The stratosphere, which is usually about –50 degrees Fahrenheit, had dropped to –80 degrees by 1999. Now, it is running at an average of –100 degrees. Meanwhile, in the troposphere, near the surface, temperatures have risen steadily, especially in the Arctic. As predicted by global warming models, moderate temperature increases in the mid latitudes have translated into dramatic increases in the far north. Temperatures in Phoenix have risen only 2 degrees above normal, but at the North Pole, the increase is 8 degrees.

Heat this extreme near the surface coupled with cold this intense in the stratosphere has not been seen before by meteorologists and is not factored into the models used to predict weather. For this reason, the superstorm is a meteorological surprise.

Indeed, it is so warm on the surface that Hokkaido, Japan's northern island, no longer

sees snow at all. But for an occasional dusting, New York has not seen it in the past few
years. That doesn't mean that there hasn't
been rain. It seems to rain all the time these
days. American scientists have even figured
out why weekends are wetter than weekdays.
It's because air pollution builds toward the
weekend, then combines with water vapor in
the air to form clouds that gather on Friday,
spilling their rain on Saturday and Sunday.

December slides into January, and the
weather grows colder. There's something of a
cold snap across Canada, and some northern
lakes freeze. In Kansas City, skies are clear
and New Year's Eve is a night full of stars
and cool north breezes. In Greenland,
though, things are different. Things are
strange.

A powerful southern flow of air has
appeared, and the Greenland glacier, already
greatly reduced, is once again sending vast
quantities of ice down the fjords. As the
warm air begins to pool near the surface,
temperatures rise, not just in Greenland, but
throughout the Arctic. Ice already weakened

from years of high air temperatures begins to melt. Great columns of water vapor rise from the much diminished polar cap, which contains huge patches of open sea even in what should be the depths of winter.

Warm air comes rushing up the coast, and while New York is enjoying yet another shirtsleeve January, the great storm begins to gather over Baffin Bay. But it isn't the only storm. Conditions are the same all across the southern Arctic. With the polar cap as diminished as it is, the flow of cold air pouring southward is not as strong as it should be, and warm air from the lower latitudes keeps coming up, raising temperatures higher and higher. In an area where it should only snow at this time of year, it begins to rain. The resultant Arctic melt releases unprecedented freshwater into the ocean and water vapor into the air. A fearsome cold wave now grips the surface, and people laugh ruefully about all the foolish global warming warnings, not realizing that the cold has actually been caused by the warming phenomenon.

At this point, surface temperatures above the Arctic Circle are well below zero. Above the surface air, the tropical flow is still pumping warm air northward. Not only is that happening, but extreme cold air from the stratosphere is being drawn downward by the increasing power of the circulation.

The storm first becomes visible on satellite images as a series of unconnected swirls of cloud, then more structural storms begin to form. They develop along a line just above the Arctic Circle, reaching their greatest circulation in areas that are over open water.

Supercells begin erupting inside the storm complexes, as the Arctic chill continues to collide with the powerful temperate flow from the south. This happens in an atmosphere that has become supersaturated with water vapor from all the melting snow and ice.

A tornado rumbles through six Siberian villages, and Juneau is hit by a storm that has eighty-mile-an-hour winds and leaves the streets buried in hail three feet deep.

Despite the lateness of the season, tropical depressions still stretch across the South Atlantic like a string of deadly pearls.

Satellites are beginning to reveal a new phenomenon: vast areas of the planet are being covered by cloud—and it is thickest in the Arctic.

For the moment, the warming trend has won. The frost line retreats northward as water vapor continues to pour into the atmosphere. But the northern seas are once again flooding with freshwater from melt. A heat wave drives temperatures in Toronto up to 78 degrees Fahrenheit, and Saint Petersburg experiences a bizarre combination of endless night and sweltering temperatures. In Moscow, even the hordes of unshod children are comfortable.

By February 5, clouds of water vapor are rising off snow and ice all the way to the North Pole, and the only stable cover ice is in patches within a hundred miles of the pole itself.

Clouds now cover the entire Northern Hemisphere as billions of tons of ice and snow are converted into water and water vapor. In the Arctic, the long nights are as black as death, and when thin light comes, the clouds can be seen to be rushing from west to east. On the ground, there is an eerie silence.

The Weather Channel is now the most watched television channel in the world— when satellite communications are not disrupted.

As the cloud cover thickens over the Arctic, temperatures stop rising. For a time, this feels more normal. But it's too early in the spring for there to be any natural warming, so the situation soon becomes unprecedented: Without the North Atlantic Current to moderate things, there is nothing to prevent cold air aloft from surging downward into the hot, wet atmosphere below.

The tropical flow being generated by the tremendous surface area of overheated seawater to the south is extremely strong. Storms of unprecedented power begin to build along the frontal lines as all this energy struggles to find balance.

But it cannot find balance. There are now disturbances from the Aleutians to Hawaii, from Sevastopol to Minsk, and over the whole of the Canadian Arctic.

Privately, meteorologists at the National Weather Center are watching in confused awe. This is the wildest weather in living memory. A tornado smashes into Warsaw. Venice is flooded by a wind-driven storm surge that comes blasting up the Adriatic. Hundred-twenty-mile-an-hour gales wrack southern England. The great floodgates in the mouth of the Thames have to be closed. In the Netherlands, a flood emergency is declared as tidal waters inundate dike after dike. Paris experiences a ferocious electrical storm that kills a dozen people. Near Kansas City, a tornado is observed on the ground and on radar that remains organized for three hours. Nine hundred lives are lost in forty trailer parks.

Despite the efforts being made to prevent it from acting, the National Oceanic and Atmospheric Administration informs the Federal Emergency Management

*Administration that it should expect exten-
sive storm damage nationwide. The Red
Cross is finally warned that there is a dangerous situation building. The National
Security Council, at last made aware of the
situation, demands answers from NOAA, but
there are no answers.*

*Winds gusting to a hundred miles an
hour sweep across New England, bringing
with them ice-flecked rain that falls in torrents. Beneath the black skies, temperatures
that were abnormally high now begin to
fall. The clouds are reflecting so much sunlight that even the overheated Tropics cannot
provide enough energy to keep them up.*

*The extraordinary energy exchange that
has developed between an Arctic fed by
extremely cold stratospheric air and the
heat-ridden Tropics begins to resolve into a
series of massive cyclonic storms that sweep
around the Arctic like swaying dervish
dancers. Some of the storms now contain
sustained winds in excess of seventy miles
an hour, with gusts as high as a hundred
and twenty. Hundreds of millions of dead*

and weakened trees are felled, transforming
what had been standing timber, or the
remains of it, into much flatter terrain.
Unhindered now, the wind screams across the
crying land.

Surface temperatures drop further. Winds
rise higher.

Refugees start appearing in British
Columbia with tales of sleet storms so heavy
that buildings are collapsing and roads are
becoming impassable. Soon, though—too
soon—the flow of refugees drops. Then it
stops.

Residents of Quebec, then Ottowa, then
Toronto begin to move south. In western
Canada, there is steady movement south into
the United States. Plans to open the border
are scrapped. The governor of Maine activates
the National Guard.

Just a few days, however, brings a deterioration in conditions so dramatic that issues
like that cease to have meaning. Hundred-
twenty-mile-an-hour winds roar across the
Great Lakes, bringing with them cold rain
mixed with snow. The cloud tops reach

higher than ever before recorded, penetrating deep into the stratosphere. As the moisture-laden air of the storms passes through this area of extreme cold, it becomes supercooled and goes plunging down toward the surface twelve miles below.

Half a dozen storms begin to organize into larger systems. And now a new phenomenon enters the picture: The air that is being frozen to extremely low temperatures in the transstratospheric cloud tops is reaching the ground so fast that it has no time to warm. Anything these columns of death touch is frozen solid within minutes. Towns and villages across the whole of the Arctic cease to communicate. There is no way to reach them, given that land movement is impossible and nothing can fly. Satellites reveal no information about what is happening beneath the cloud cover, but the victims— some of them frozen so quickly that their dinners are still in their mouths—will not be found again for thousands of years. Like the mammoths who preceded them in the last storm, their remains will suggest to the future that something strange and terrible happened, but the memory of the cataclysm will have died with them.

As the storms continue to organize, the tropical flow itself becomes dangerous. Sustained winds in excess of 140 miles an hour begin along the face of the Rockies from the south, destroying all human habitation in their path. Airborne activity is now impossible north of a line from Denver to Richmond.

A special meeting is convened at the White House, but NOAA officials from Colorado must attend by telephone, as travel is too slow and hazardous. Refugees from the north have panicked people in the central U.S., and a wave of southward migration threatens to disrupt the economy and overwhelm all services.

During the meeting, wind gusts in Washington exceed a hundred miles an hour. Russia is no longer available by telephone. Sweden, Norway, and Finland are all reporting unprecedented spring snows accompanied by blizzard conditions and winds of 130 miles an hour.

The royal family has retreated to Balmoral in Scotland and been trapped there by a crushing ice storm that has immobilized the entire area. They will never be seen alive again.

The president declares a state of national emergency and imposes martial law. He now calls out the National Guard, but events have moved so quickly that only a few units can manage to get organized.

Highways are packed with cars moving south—or, more often, not moving. The NOAA facility in Colorado goes out of commission due to wind damage.

Two days after the meeting in the White House, the government begins to make plans to move out from under the storm area. But it's far too late. The president and his entourage, forced to use the roads like everybody else, join the stalled mass of traffic that jams highways from Virginia to Texas.

Over water, sustained wind velocities now exceed two hundred miles an hour. The storm is now fully organized, a single, massive cyclone. It functions just like any storm, but on a scale so much larger that the possibility has never been considered. Its circulation extends from the tropic of Cancer to the Arctic Circle.

Stockholm, Helsinki, Moscow, Saint Petersburg, Edinburgh, Toronto, Vancouver, and Fairbanks are among the large cities that are no longer communicating in any way.

The satellite relay station at Pine Gap in Australia is reporting problems communicating with California due to weather extremes.

Pine Gap ceases to receive communications from CIA headquarters at Langley, Virginia. Then it ceases to receive communications from the control center in northern California. American interests in the Southern Hemisphere are isolated.

China has joined Russia in a chorus of silence. Stations still recording indicated sustained winds across the northern Pacific upwards of 230 miles an hour. Farther north, wind speeds may be even higher, approaching the maximum possible on earth, but meteorologists can only guess.

The Weather Channel, operating from Miami, continues to broadcast what information it can. Florida, Alabama, Mississippi, Louisiana, and Texas have received 30 million refugees in ten days. Having lost communication with federal authorities, the governors of these states have formed an ad hoc alliance. Continental Army Command has agreed to report to the emergency governing committee. State martial law has been declared, and looters are being shot on sight in hard-hit cities like Dallas at the southern extreme of the storm zone, which is both severely damaged and choked with 2 million desperate refugees.

Incredibly, the storm continues to grow. Outside of the storm area, air pressure drops lower than thought possible. The planet's atmosphere itself is being distorted—and this, in the end, is what finally unbalances the weather system and causes it to begin to break up.

As this happens, the storm turns loose the water vapor it has been carrying, dropping it in the form of a vast, continent-girding snowfall.

This process finally equalizes the energy, and the storm slowly winds down. It takes two weeks, though, before so much as a shaft of sunlight reaches the surface of what had been the richest and most developed part of the world.

After the storm, the landscape has completely changed. There is not a tree standing north of Oklahoma. The land is a glaring sheet of ice. At the Canadian border, the ice is ten feet thick. Across the tundra, it's forty feet thick. The northern polar cap has reappeared and seen from space it appears to have grown threefold in just a few weeks.

This is deceptive, however: this is all new ice, very loosely compressed. As March turns to April, its southern reaches begin to melt. A month later, the Mississippi is ten miles wide in places. New Orleans is inundated. The Delta is destroyed.

Over the summer, the most massive flood that mankind has ever seen—at least in historical times—takes place. Those areas that survived the storm must now deal with its aftermath.

Across low-lying areas of Alabama, Louisiana, Mississippi, and Texas, a huge lake develops. It will be a short-lived phenomenon, lasting only a few years and declining steadily throughout that period, but it will also drown a substantial number of the livestock remaining in the United States. Whole cities will be inundated.

Farther north, the air temperature remains low. This is because the ice itself is so reflective, and also because the winds speeding across the North Atlantic have caused so much evaporation that the salinity of the water has risen again, tripping the system of currents even further out of balance.

The North Atlantic Current now returns southward off New York instead of Greenland. The Gulf Stream no longer goes north, bringing mild winters to Europe. Instead, it ends near the Bay of Biscay, near France and Spain. England now has a climate similar to Lapland's before the storm. Next winter, the frost line in the United States will extend all the way to central Florida.

The civilization of the northern peoples, which has been of such central importance to mankind for thousands of years, has taken a truly astonishing blow. Underneath all that ice, being ravaged by those howling winds, a billion people have died in what was the planet's most productive, most educated, and most civilized region.

A shattered United States, now reduced to a strip of still-intact states bordering the Atlantic and the Gulf of Mexico, begins to fail in fundamental ways. The states, lacking a central authority, fall away into independent entities.

Western civilization as we know it passes into history.

15

Distant Thunder

~~~~~~~~~~~~~~~~~~~~~~~~~~~~~~~~~

FIVE ENORMOUS EXTINCTIONS HAVE BEEN IDEN-
tified in the past, each with very high numbers of
species involved. These are named for the geologic
periods that they closed: the Ordovician extinction, which destroyed eighty-five percent of all
species 439 million years ago; then the Devonian,
Permian, and Triassic events; and then the Cretaceous extinction, which destroyed the dinosaurs.

Each of these had unique characteristics, and it
is not clear what triggered most of them, although
comet or asteroid impacts brought about the climaxes of a number of them.

The Permian extinction, with its ninety-five
percent kill rate, was the worst of them, and there
is some evidence that it was started by extreme

global warming that developed over a long period. Like the other events, the climax was preceded by a drop in sea levels approximately 2 million years prior, which began a long unwinding of the biosphere, resulting in an ever-growing number of species being killed off.

During the Permian climax, thecodonts, one of the ancestors of modern mammals, began to actually tunnel into the soil in their desperate effort to escape the heat. Their fossils are characteristically found in the Karroo Desert of South Africa, entombed in the tunnels that they dug and died in a quarter of a billion years ago.

The Cretaceous extinction, which killed off the dinosaurs, involved the death of about seventy-five percent of all species then living. While there is growing evidence that some dinosaurs evolved into birds, so many of these creatures disappeared that the entire genus was effectively wiped out, at least in its characteristic form.

What were conditions like during these mass extinctions? There is always an assumption that they took place suddenly, that the extinction event involved a dramatic breakdown of some kind. But the events seem to have primarily been caused by a combination of geologic and climatic changes that only slowly developed to crisis proportions. For example, the Permian extinction appears to have come about when geologic changes blocked the flow of ocean currents, and the oceans, then the air, became stagnant. This process would have taken hundreds of thousands of years, during which time species extinctions would have occurred at an ever rising rate, coming to a climax when ocean heat reached a level that killed plankton, and this fundamental link in the food chain was broken.

That the climax was sudden is made clear from the many fish fossils found from this event, often in schools, which confirms that the animals all died together—probably, in this case, of exhaustion caused by overheating, lack of oxygen, and starvation.

The final end of the dinosaurs came as a result of an amazing upheaval that was caused by a large nonterrestrial object hitting what is now the Yucatan in Mexico. This collision spewed huge amounts of debris into the atmosphere, while at the same time setting fires over millions of acres, so that the entire planet became choked with smoke even as temperatures dropped radically due to the lack of sunlight reaching the earth's surface.

Under these circumstances, extraordinary storms would have developed, but they are not related to the kind of superstorm that can apparently climax the warming periods between ice ages.

In any case, this final event was so destructive that what had been a slow, persistent decline became, as in the Permian extinction, a tremendous die-off right at the end. The actual amount of time is debated, but it appears that the greatest carnage took place over about a ten- to twenty-thousand-year period—about as long as the climax of the extinction event we are in now has been under way.

What kind of a picture does the world present after an extinction climaxes? The best available fossil record comes from the dinosaurs, and it is a sobering one. Not only were they completely destroyed, most other species suffered as well. There were huge losses among sea creatures of all kinds, among mammals, among plants. The extinction event continued for thousands of years after the impact. Even a thousand years later, the atmosphere would still have been opaque from the debris that had been thrown into the air and the pollutants that were being spewed out of volcanoes that had become active because of the shock of the strike. This volcanism put far more pollution into the atmosphere than human industrial activity has. In fact, so much sulfur dioxide poured out that sulfuric acid rain resulted, dissolving everything it touched.

In the end, the land was covered in hardy scrub plants and the oceans were almost empty of life. Animal populations were very scarce, except for insects and fast-proliferating species like small mammals, which were also adaptable and capable of eating almost anything they could swallow.

Dire conditions persisted for hundreds of thousands of years after the impact. Life came back, but only very slowly. In all, it seems to take between 5 and 10 million years for life to return to the level of diversity that was present before an extinction event. Had mankind been around during such a period, we probably would not have realized that we were living in a postextinction world any more than we can readily see now that our entire history has taken place during an extinction event.

The present mass extinction did not begin because of man. The sudden overgrowth of human population is, rather, a reflection of climatic imbalances that have temporarily favored our species over others that are less intelligent and flexible. Our exploding population is part of the extinction climax.

If present rates of extinction continue, it would appear that the present event will climax with a loss of about two-thirds of all species—not as bad as the Permian extinction, but somewhat worse than the Cretaceous.

This will mean that the earth will again be left with a very small number of hardy creatures still alive—things such as rats, weeds, roaches, mosquitoes, and such. And man? Perhaps, but our large size is not a plus. During extinctions, the larger species are generally the most vulnerable. Already, the present extinction event has killed many of the large species of the planet, which filled every continent as recently as ten thousand years ago. If the normal pattern is followed, the next wave of the event will concentrate on the second tier of large animals, of which mankind is one. In the past, highly adapted, high-population species with rigidly established dietary requirements have been most at risk, so our ability to eat many different things may help us. But our concentration in urban centers that render so many of us relatively unable to cope with more rigorous living conditions means that we have fallen into the old trap of habitat inflexibility that makes large species so vulnerable.

The structure of extinction events appears to follow a rough pattern, but it must be remembered that the fossil record is so difficult to read and so incomplete that there is always room for controversy. However, the period before, during, and after the event that destroyed the dinosaurs has been studied extensively, and it can probably serve as a template for other events.

About 2.5 million years before the impacting asteroid or comet ravaged what remained of the dinosaurs, the extinction event began. As at the beginning of the Permian extinction hundreds of millions of years before, there was a dramatic global temperature change, followed by a drop in sea levels. Coral reefs and bottom-dwelling marine creatures were extinguished. There were climate changes that caused a wave of extinctions among plants. The climate became drier and colder. Huge jungles dried up, and grasslands appeared.

When that happened, a process took place that would be repeated many millions of years later, when it would have something to do with us. Racing across those plains, there came bright, quick dinosaurs, the brightest a creature called *Struthiomimus.* These plains-dwelling predators probably were fast and smart and may have hunted in organized packs—not unlike the bright land-dwelling apes that appeared at the beginning of the present mass extinction . . . the clever, quick little apes that have become the human beings of today, that must now cope with the climax of the extinction event that brought us about in the first place.

Hopefully, we will be more successful than our predecessor species—we who also emerged with the drying up of great jungles onto dangerous plains, where speed and brilliance combined to make us an excellent competitor.

By the time the comet that killed the dinosaurs struck, their world had—like ours has now—already been in upheaval for millions of years. Like us, *Struthiomimus* lived out its entire history inside the context of an ongoing period of extinction. *Struthiomimus* was eventually overtaken by it, and the earlier brilliant predator became extinct.

The current extinction process, as before, has not been continuous, but rather has come in pulses. Each time, life has adapted to the changes and diversity of species has risen again. But this time, the number of species becoming extinct has gone so high and the extinctions are proceeding so quickly that it is unlikely that there will be a quick return to diversity after the present process plays itself out.

During the past twenty thousand years, we have entered the same kind of intensive extinction climax that closed the era of the dinosaurs altogether. As happened then, the chemical content of the atmosphere is changing rapidly. But in those days, the disaster was caused by burning continental forests and volcanic eruptions. Nowadays, it is caused by chemical factories and the burning of fossil fuels.

No matter, though; the effect is the same. The rate of extinction now is about what it was after the object that killed the dinosaurs had struck.

It took five to ten million years for the earth to regain animal populations that even approached the dinosaur era in numbers. Especially in the oceans, there were many species that started and failed, many evolutionary misfires. It was not until 35 million years had passed that earth again bore a stable population of creatures as diverse and highly evolved as had existed before the extinction.

It is almost impossible for us to think in such vast epochs of time. But it is only over such stretches of eons that the history of life on earth begins to reveal its larger meanings.

Just under 3 million years ago, something happened that disturbed a period of nearly 60 million years of relative climatic stability and gradual species change. The disturbance was radical enough to identify this era of ours as an extinction event.

What happened to cause this change? Why are

we in the remarkable situation of coming to understand our peril almost at the exact moment that it is overtaking us?

To get the answers to these questions, we must go back to that era, when the long series of ice ages that have marked our epoch began.

16

Panic Stage

FROM: BOB MARTIN, NATIONAL SEVERE STORMS
LABORATORY, U.S. NATIONAL OCEANIC
AND ATMOSPHERIC ADMINISTRATION

TO: ALEX RICH, THE HADLEY CENTRE FOR
CLIMATE PREDICTION AND RESEARCH,
U.K. METEOROLOGICAL OFFICE

ALEX, I'M DAMN SORRY I HAVEN'T WRITTEN
BEFORE. IT'S BEEN HELL HERE, OBVIOUSLY. WHAT
PREPARATIONS ARE YOU MAKING? AND I MEAN PER-
SONALLY. WHAT ABOUT JANET AND THE KIDS? I

THINK PORTUGAL—SOUTHERN PORTUGAL. OF COURSE, IT'S GOING TO BE JAMMED. JESUS, I'VE SENT MARTIE AND OUR KIDS TO TEXAS.

IT'S UGLY, GETTING THEM OUT BEFORE THE PUBLIC KNOWS. BUT WHAT CAN WE DO? YOU HAVE TO THINK OF YOUR OWN FLESH AND BLOOD. WE ARE HAVING SO DAMN MANY MEETINGS YOU CAN'T WORK. THE WHITE HOUSE IS SCREAMING, THE NATIONAL SECURITY COUNCIL IS SCREAMING, FEMA—FEDERAL EMERGENCY MANAGEMENT, REMEMBER, BUT WHY WOULD YOU??—ANYWAY, IT'S SCREAMING, AND NOBODY HAS—I MEAN, NOT THE LEAST DAMN IDEA.

HAVE YOU SEEN THAT THING THAT WENT UP OVER BAFFIN?—MY GOD, I THINK THERE MIGHT BE A 100% KILL RATE UP THERE. WAIT UNTIL THAT HITS THE PRESS, MAN. YOU WERE LOOKING AT POCKETS WHERE THE TEMP DROPPED A HUNDRED DEGREES IN MINUTES—I MEAN, JESUS. WHO KNEW THE WEATHER COULD DO THAT?

I WORRY ABOUT YOU, MAN. I WANT A REPLY TO THIS. THE U.K. IS A DEADER. YOU WILL NOT MAKE IT. THE FLOW—WE'RE GETTING THIS WHOLE THING FED AS A SINGLE STORM, I THINK—THE FLOW, ANYWAY—SORRY I'M DISJOINTED. ABOUT

FIFTY PEOPLE ARE SCREAMING AT ME, AND I'VE
BEEN LIVING ON COFFEE EXCLUSIVELY FOR SIX DAYS
SO I'M WIRED. STARTED SMOKING AGAIN—WHY
NOT? HA HA HA (NOT A JOKE, SORRY)—THE FLOW
IS UNBELIEVABLE. THERE IS A PLANETARY
STRUCTURE TO THIS THING.

YOU ARE GOING TO SEE MASSIVE SNOWFALL,
WIND. WOW. I MEAN, I AM TALKING MAYBE THREE,
FOUR HUNDRED INCHES OVER SAY A COUPLE-WEEK
PERIOD. THEN A SOLID FREEZE. YOU CAN'T LIVE.
THE U.K. IS A DEADER. DID I SAY THAT? WELL, I'M
TRYING TO IMPRESS YOU.

THIS IS NOT ABOUT THOSE YOU DON'T KNOW,
MAN. THIS IS ABOUT THOSE YOU LOVE. BUY GOLD.
GOLD COINS. WHATEVER THE HELL THEY SELL.
NOTICE, THE PRICE IS STARTING UP. WELL, IT WILL
GO CRAZY. YOU GET UP OFF YOUR DUFF AND GO TO
WHEREVER THE HELL THEY SELL THEM OVER THERE
AND YOU BUY GOLD COINS RIGHT NOW, MAN.

THEN GO HOME AND GET YOUR FAMILY AND
GO SOUTH. BUT BE CAREFUL. I WORRY THAT
YOU'LL BE TARGETS, YOU BRITS. PORTUGAL MIGHT
BE THE SAFEST. THEY ARE SUCH SWEET PEOPLE
THERE. REMEMBER THAT—GOD, IT WAS TEN DAMN
YEARS AGO, BUD—BUT IF NOT, THEN TRY GIBRAL-

TAR. TOO BAD NO COLONIES LEFT, MAN. YOU PEO-
PLE SURE COULD USE THEM NOW.

YOU CAN SEE HOW IT'S GROWING. EVEN WITH
JUST THE SAT DATA. WE HAVE ROTTEN DATA OVER-
ALL. WINDSPEED, ETC. YOU CAN'T PENETRATE WITH
A PLANE. THE ICING IS TOO MUCH. SHIPS—WE LOST
GUYS ON A SHIP DROPPING BUOYS. IT WAS AWFUL,
THEY STARTED CALLING ON THEIR CELL PHONES.
JESUS.

OH, CHRIST, HOW CAN I SAY THIS? TO HELL
WITH DUTY. JUMP SHIP. THERE, I DID WHAT I
GOTTA DO. BUT I KNOW YOU WON'T. ME, NEITHER.
I'M HERE TILL THE DAMN ROOF COMES DOWN ON
MY BALD DAMN HEAD.

GOD BLESS YOU, ALEX, AND EVERYBODY IN
THE MET FROM ME AND ALL OF US AT NSSL.
WEATHER PEOPLE ARE FAMILY—BOY, DO YOU FEEL
THAT RIGHT NOW. MAY WE MAKE IT TO BETTER
DAYS, BROTHER-IN-LAW.

BOB

17

Storm Signals

~~~~~~~~~~~~~~~~~~~~~~~~~~~~~~~~~~~~~~

THREE MILLION YEARS AGO, WHEN THE GEOLOGIC
epoch known as the Pliocene was ending, earth
was a very different sort of place from what it has
been since then. Looked at from above, the continents would at first have seemed a lot like they are
now. But there were no polar caps, and—of utmost
importance—there wasn't any land connecting
North and South America. Because of this, the climate was totally different from what it became
when the Central American land bridge appeared.

The Americas were separated for so long before
that happened that entirely different species
evolved on the two continents.

If a time traveler were to go back to the middle
of the Pliocene, he would find a world rich with

life, full of diverse forms, and healthy in ways that we have never known.

In fact, mankind has not lived in such a time, when the world was in the profound state of climatic balance that it enjoyed then and that it usually enjoys. The brief period of stability that we have experienced during our recorded history has come in the context of a great cataclysm. It is only one of the many pauses that have taken place between grueling eons of ice.

For the past 3 million years, earth has been in an agonizing cycle of alternating ice ages and brief warming periods. Looking back to the world before this period is like looking at a vision of a distant Eden.

The recovery from the catastrophe that destroyed the dinosaurs had been slow, but by the end of the Paleocene epoch, 10 million years after that event, earth was teeming again. Huge, rhinolike creatures with six blunt horns called *Uintatherium* roamed wide areas of savanna, chased by peculiar predators with hooves and pointed teeth, like—of all things—carnivorous horses.

Forty million years ago, the earth started an immensely long process of cooling. The change ended the warm, placid Eocene epoch and began the cooler, drier Oligocene. It was very gradual. There were no great upheavals. But by eleven million years ago, the difference was profound.

During the Oligocene epoch, more modern creatures began to appear. The exuberant sense of the absurd that seems to characterize creatures in the early epochs of a great evolutionary period gave way to more efficient, successful animals. One wonders if some of the earlier creatures—ludicrous monstrosities like the gigantic *Baluchitherium,* an immense, ungainly mass of archaic mammal flesh—can have had endurable lives, given the sheer amount of inconvenience that must have been involved in their existence. The early years of the Age of Mammals must have echoed with disappointed roars, bleats, honks, and rumbles as such creatures strove to make their impossible bodies work.

But as grasslands steadily replaced jungles, more efficient species began to appear—forms so stable, indeed, that their shapes are still familiar to us in the animal world of today.

The cooling continued. This slow, steady change is believed to have been due to a long-term decline in the sun's heat, because few other causes could have been so consistently present over such a long period.

Then, about 6 million years ago, during the Pliocene, an event rare in earth's history began to occur. The Antarctic continent, which had slowly drifted into the southern polar area, started forming an ice cap. While sea ice often covered polar waters, it was rare for a landmass to end up in such a position.

As the continent began to ice over, ocean levels dropped so low worldwide that the Mediterranean became an inland sea and then dried up. The entire area was in a state of profound drought, otherwise entering rivers would have fed the isolated sea, stabilizing its decline at some point.

At this same time, the African forests in which our primate ancestors lived collapsed, and the long journey toward man began.

The creature that had to cope with this change was probably a proto-ape called *Paranthropus robustus,* which began to spread across Africa.

It was at this time, also, that a period of geologic unrest started. Volcanism began to spread across the world. There were upheavals, although nothing serious enough to cause another mass extinction.

This volcanism brought a geologic event of enormous importance. In fact, if we think that the advent of man was the most important thing that ever happened to our planet, then this was among the most crucial of all geologic changes.

As a result of all the volcanic and seismic activity, Central America rose out of the sea. It blocked the all-important flow of ocean currents around the equator that had served to stabilize earth's climate for millions of years.

The first effect of this change was that North American predators such as the saber-toothed tiger *Smilodon,* one of the most fearsome cats ever to stalk the earth, arrived in South America and proceeded to go on an eating spree that must have lasted some hundreds of thousands of years. *Smilodon* and other

North American predators were probably responsible for the disappearance of a number of exceptionally clumsy South American herbivores, such as the anteater *Nothrotherium*, which must have moved about with all the agility of a giant land tortoise, but without a protective shell.

This alteration of the earth's geography contributed to what would soon become the biggest change in climate that had happened in sixty million years. It was the beginning of the great extinction event during which we have appeared.

Our branch of the family of primates has reacted to the continual pressure of an environment in chaos by becoming ever more adaptive and intelligent. In Africa, *Paranthropus* disappeared, replaced by a series of more advanced primates. By 3 million years ago, *Australopithecus africanus* began to roam the spreading veldt. *Australopithecus* was a hunter and likely tool-user, and is thought to be the first human ancestor. He probably hunted in groups and possessed a social structure. He was smarter than the modern chimpanzee. The world he lived in was a hard one, plagued by drought and wracked with change. All around him, species were dying.

Because of the Central American land bridge, ocean currents were forced into a north-south flow. A new system of heat exchange between the Tropics and the Arctic began. It brought with it weather of a kind that had not been seen on earth in a very long time.

It must have caused a huge die-off of forests, because the carbon dioxide balance of the air changed noticeably. CO_2 levels began to fall. Over a million-year period, they were cut in half, and kept falling. This meant that the atmosphere was less able to hold heat. At the same time, the diverted ocean currents resulted in a much more violent and variable climate. The differences between seasons became dramatic, and the huge temperate forests that reached into the Arctic began to give way to a taiga (subarctic forest) characterized by stands of pine.

Winters grew longer. The North and South Poles were now covered by ice year-round. In the north, the ice crept south from a number of centers, the most important one in Hudson's Bay, crushing what remained of the ancient temperate forest as it came, disrupting animal life, destroying food supplies, putting every living creature under pressure to survive. Ahead of the ice, the taiga spread for hundreds of thousands of square miles. In this taiga, bears large enough to tear a human being in half were the dominant predatory species.

Blizzards and hurricanes became commonplace, and tornadoes must have marched the plains. At this time, the sort of superstorm that may mark the change between ice ages and warming periods would not have occurred, but it probably became a rare but devastating feature of our climate in the near future.

For the past 3 million years, the ice cover has waxed and waned, and every time the ice has reappeared, the extinction rate has soared. Whole species vanish in the snows and droughts every time the glaciers come crashing south again. Ice does not spread north of the Antarctic continent because it is surrounded by oceans, although climatic changes in the Southern Hemisphere are also extreme, and local glaciers in New Zealand and South America grow dramatically during cold periods.

The world is now in a state of spasmodic alternation between long ice ages and short warm periods. During this time, all creatures have been put under extreme survival pressure. Around fifty thousand years ago, old species began dying out faster than new ones emerged.

During this time, the hominids have continued a process of rapid evolution, challenged by the harsh and ever-changing conditions. There have been a series of at least six hominid species during this geologically short period, each more of a tool-user than the last. *Homo erectus,* who preceded *Homo sapiens,* was considerably smarter than his predecessors. He not only used tools, but, judging from the recent finds in Germany that were discussed earlier, was an expert carver of spears and—from the Indonesian evidence—was even capable of navigating the oceans out of sight of land.

From our beginning, we have used our abilities of mind to respond to the environmental pressures our age has put us under. To survive, we have learned all manner of craft, from chipping flint and sewing hides to making gasoline and insulation. We have become scholars and architects and engineers, developed complex societies, and done it all in response to the pressures of an environment that has betrayed us many times in the past and will again . . . many times.

Our problem right now is that we have been too successful, with the result that our success has, itself, become a significant contributor to our danger. Our intelligence has enabled us to proliferate to the point of imbalance, and imbalances do not last long in nature. By the middle of the next century, even if population growth remains at its present moderate rate, the planet will be called upon to feed, clothe, and shelter two people for every one alive today.

Although social influences are reducing population growth, the number of human beings is now so huge that even one percent growth per year is a disaster. Worse, the more prosperous we become the more each individual consumes—so not only are there too many of us, we weigh more and more heavily on the earth.

Mankind is the *Struthiomimus* of the present extinction climax, reacting to the extraordinary pressure by evolving intelligence. But humankind has been far more successful in this regard than any known species, past or present. Brilliant though he was for a dinosaur, *Struthiomimus* was probably about as smart as a fox terrier.

In its torment, our epoch has evolved something astonishing, a creature that is at once genius and monster, the first animal ever to find itself in the ironic position of actually worsening, with the intensity of a natural force, the very process of generalized extinction that threatens to destroy it. As our numbers have increased, we have gone beyond destroying other species through excessive predation. Now we are destroying them as part of the process of providing ourselves with more space. They are being pushed aside by mankind as if by a new environment that they cannot inhabit, a spreading desert, even a lava flow.

Man's great population centers function like the lava flats of the dinosaur era, covering the land with a hard, rocky surface. Like the continental fires that followed the impact of the comet that destroyed the dinosaurs, we are emitting vast quantities of combustion products into the atmosphere.

So far, we have not produced quite as much smoke or quite as much acid rain or quite as heavy a load of pollutants, but we have certainly caused every bit as much extinction. Humankind *is* the catastrophe that is bringing on the climax of the present extinction event.

The logical question is, What can we do to transform ourselves from a natural disaster into a blessing? Just as nothing could have interrupted the effect of the comet on the dinosaurs, we cannot end our own effect on ourselves and our fellow creatures. However, we can develop strategies that might stave off the disaster long enough to enable us to substitute effective human planning for the automatic natural processes that govern—and threaten—our world now.

Despite this, though, we are, along with everything else on earth, subject to the great cycle that controls the planet. Human activity alone has not created the change of climate that is coming. This same sort of change has happened at the end of every warming period ever since the Central American land bridge changed the ocean currents and began the cycle of overheating and overcooling in which we are trapped.

But how is it that earth has changed, in just under 3 million years, from being a richly alive planet with a steady-state environment to the fantastic situation that now pertains: a world in the grip of a deadly natural cycle that is threatening the lives of billions of the most brilliant of its creatures?

To find out, we must explore a little more deeply the most recent of its climate changes, the sudden return of cold that has haunted the pages of this book, the catastrophe that unfolded about 8,000 years ago. If there are superstorms, then one may well have struck at that time. It wasn't quite powerful enough to bring the ice back, but we might not be so lucky the next time.

18

Does It Happen?

SUDDEN CLIMATE CHANGE IS AN ESTABLISHED scientific reality. But the superstorm is another matter entirely. Nobody has been able to study what happens during a period of sudden change, because we can only measure its effects, not observe the event itself.

As we saw in Chapter 7, however, when read one way the fossil record tells a story of sudden and devastating upheaval and the strange, horrible deaths of creatures.

In looking back into the recent geologic past, we find that we are dealing with three separate but almost certainly related climatic events. The first took place approximately seventeen thousand years ago, when earth's climate started heating up and the glaciers went into a rapid melting phase. The second occurred twelve thousand five hundred years ago (10,500 B.C.), when there was an even more dramatic increase in temperatures. Four thousand five hundred years later, or about eight thousand years ago, a sudden period of cooling took place, followed by two hundred years of much colder temperatures.

If there was a proto-civilization, this may be when it was destroyed. Unfortunately, it is impossible to claim that all of the structures we have been discussing that were created with unknown engineering techniques were built before that time. The only ones upon which such dating can be placed with any degree of conviction are the Sphinx and the sunken ruin off Yonaguni—if it is indeed a ruin.

However, there was definitely a climatic event then that appears similar to our present situation.

In October of 1998, a study of new ice cores from the Antarctic confirmed earlier research from Greenland that suggested an astonishingly rapid advance in temperatures around 10,000 B.C. An article published in the journal *Science* indicated that temperatures in the Antarctic rose 20 degrees Fahrenheit over a very short time, during the same period that Arctic temperatures rose an incredible 59 degrees. This resolved what had been an open question in science: whether the temperature increase seen from the Greenland cores was local or part of a worldwide change. Dr. James White, a

University of Colorado climatologist, said that the findings "throw a monkey wrench into paleoclimate research and rearrange our thinking about climate change at that time."

This massive change would have caused upheavals on a tremendous scale, unlike anything seen since. To accomplish it would have taken a gigantic amount of energy. For example, for sunlight alone to account for the heating that preceded the cold snap, the sun would have had to have put out around ten percent more energy than it had previously. But there is no evidence that the sun could have done this, and there is a certain amount of evidence that an output increase on this scale could have occurred only if there was a destabilization of the structure of the star itself. On the contrary, according to Todd Sowers of the Lamont-Doherty Earth Observatory and Ed Brook of the University of Rhode Island, temperature fluctuations over the past 110,000 years are not related to solar output, but closely connected to the atmosphere's gas mix, specifically to the amount of methane present at any given time.

Guy Rothwell of the Southampton Oceanography Center reported in *Nature* (vol. 392, p. 277) that a massive landslide took place in the Mediterranean about twenty-two thousand years ago, when the last ice age was at its most severe. This underwater landslide caused huge amounts of methane trapped in undersea soils to escape into the atmosphere. Over half a billion tons of methane might have been released. This would have almost doubled the amount of methane in the atmosphere in just a short time.

The result? Temperatures warmed abruptly, beginning the trend that caused extensive melting of permafrost and the release of even more methane, with the result that the 120,000 year long grip of the cold was broken.

But that isn't all that happened, and the trigger that ended the last ice age doesn't interest us as much as the subsequent cooling event.

About eight thousand years ago, toward the end of the last geologic epoch, the Pleistocene, there was an upheaval. Methane levels in the atmosphere suddenly plummeted by twenty percent. Cooling followed.

Within two hundred years, however, the earlier warming trend was back in place. Subsequently, temperature and methane levels have continued to rise together, and in recent centuries carbon dioxide and other combustion-related gases emitted by human activity have intensified the effects of natural atmospheric heat retention.

We now know from the ice cores that abrupt climate change is an indisputable reality of life on earth, given its present geographic configuration. But a climate shift has not happened to 6 billion human beings before, living in a massive civilization that absolutely depends for its existence on continued climatic stability.

Before we find out how accurately we can predict the next event, let's look at the last one. Eight thousand years ago, glaciers still stretched as far south as northern Canada, but the ice cover had dwindled dramatically.

Egypt and the Middle East were a temperate zone, with a significantly greater rainfall than at present, although less than it had been just a few thousand years before. African animals such as lions still roamed the entire Mediterranean area. Southern Europe had milder winters than in many thousands of years, and tundra did not begin until the latitudes of northern Germany. The British Isles were not connected to the mainland, but Calais was almost within shouting distance of the wide marshes that stretched below the Dover Cliffs.

At that time, something happened. A sudden *decrease* in the methane content of the atmosphere of twenty percent is as astonishing as the earlier doubling. Temperatures plummeted, but only for a short period of time. During this period, however, the world appears to have been devastated. It is then that we speculate that a superstorm occurred.

The final wave of extinctions of large animals took place. During this period, the last of the mammoths died. They had followed their range northward during the warming. As temperatures rose, the forests expanded from the south, and the grasslands that they needed to survive moved ever closer to, and finally above, the Arctic Circle. By the time their final extinction took place, mammoths and mastodons were migrating north onto the tundra during the summers.

Judging from the remains of trees that were caught in it, the sudden cooling event that took place eight thousand years ago happened during the summer, when the mammoths would have been at the northern extreme of their range. Their most recent remains are generally found well above the Arctic Circle.

They have been dismissed as a piece of scientific trivia by paleontologists, who explain their deaths with the scenario that they fell into cracks in the ice and were trapped inside the frozen glaciers. As we saw earlier, however, they were caught grazing, and killed so suddenly that vegetable content was found in their stomachs, meaning that digestion stopped suddenly.

What were the fields where they died like? They contained grasses, flowering plants, and fruit trees, the remains of all of which have been found in or near the frozen animals.

The sudden freezing that killed these animals required much more than a bad storm. It required a storm that was capable of delivering unprecedented levels of extreme cold to the surface and doing it so suddenly that the animals, which were caught placidly grazing, did not even have time to look up.

Whatever was going on around them before they were killed, it was not enough to alarm them in any way at all. Frightened animals do not die while grazing. They die in flight. But these animals were not running. To all appearances, they were simply frozen solid where they stood, without enough warning to make them do more than raise their heads.

The most famous of the frozen mammoths, the Berezovka mammoth, is mounted in the Zoological Museum in Saint Petersburg, Russia. Discovered in 1901, the animal was intact except for its trunk and other areas that had been eaten by predators before the remains could be removed.

The Berezovka mammoth is preserved just as it was found. It died in the middle of a sudden movement, raising its trunk as if surprised. It had been grazing, as is confirmed by its stomach contents. It did not fall over even though it had sustained a crushed pelvis while it was alive. There has been an argument made that it fell into a hole in the tundra and froze there.

However, it is not obvious that the mammoth was capable of living in arctic conditions, despite popular lore to the contrary. The mammoth was covered with hair, not fur, and its skin lacked the type of erectile muscles that evolve in hair-covered animals who are exposed to extreme cold. In addition, the frozen mammoths are not found in isolation. A whole landscape of flora and fauna died with them, including fruit trees and plants associated, not with the tundra, but with much warmer conditions. The larvae of the warble fly have been found in the intestines of frozen mammoths, just as they are found in the intestines of tropical elephants now. The remains of horses, rhinos, and other animals found in the same areas as the mammoths were not arctic-adapted either.

As recent climatology has confirmed, the subarctic was substantially warmer eight thousand years ago and had been warm and getting warmer for around three thousand years.

The area then underwent radical freezing and has not thawed since. At the same time, a powerful freeze caused the human population of Europe to fall. Two hundred years later, the interruption of the longer term warming trend was over . . . and flood legends began to be repeated all over the world.

What appears to have happened is that the post-glacial warming trend was briefly interrupted by a violent period of cooling. Warming resumed a few hundred years later, but left the Arctic frozen. At present, this area is just now starting to thaw again.

What sort of catastrophe can possibly have done this? Perhaps it means that the land the animals were on was moved to a much higher latitude, as theorized by Dr. Charles Hapgood in his book *Earth's Shifting Crust.* Hapgood's work has been thoroughly dismissed by the rest of science, but this is probably only an emotional response by people who are uncomfortable living with the idea that such tremendous changes could happen so quickly and be driven by mechanisms that we don't understand at all.

Whether or not a shift of the crust was involved, whatever did happen most certainly included an extremely severe weather event. But this monstrous event left more than just the one cryptic sign of the deaths of the mammoths. It left the footprints of its passing all over the Northern Hemisphere.

The most telling geological evidence that extreme winds sometimes blow across the Northern Hemisphere can be seen in places like the Hawaiian Islands. The Big Island's topography, for example, reveals evidence that massive seas and severe winds sometimes come from the southwest, creating eddies that surge around to the leeward side of the island, where they are so intense that they literally scour away the cliffs that can be seen there.

These winds are apparently accompanied by volcanic activity, as lava flows can be seen on each cliff shelf, flows that have been dated to one-hundred-thousand-year intervals by the U.S. Geological Survey. No similar combinations of wind erosion and lava flows are found elsewhere in the Pacific, and if there were superstorms, this particular area would be affected, while islands farther south would not.

The storm would be a truly extraordinary event, not only because of its huge size and extreme wind velocities, but also because its rapid circulation would move supercold air down from high altitudes so fast that episodes of instantaneous freezing would occur on the surface. This sort of phenomenon might explain the sudden freezing deaths of the animals that have been found.

Three questions must be asked: What caused this to happen; is there any record of it in human myth; and is it going to happen again?

Given that billions of human lives may be at stake, these questions need to be answered.

19

Voices from the Storm

~~~~~~~~~~~~~~~~~~~~~~~~~~~~~~~~~~

LEGENDS ARE FUNNY THINGS. NOBODY CAN TELL
how old they are. Modern science, emerging out of
the Age of Reason with a built-in bias against non-
Western ways of thinking, dismisses the stories of
older cultures as flights of imagination and does little to address their origin.

But because the Bible is viewed as part of the
Western canon, its myths have been treated differently. There has been a great deal of interest in connecting biblical stories with actual, historical
events. There is even a substantial body of work,
most of it hopelessly misconceived, that tries to
reinterpret the whole nature of the past based on
what is called a "literal" interpretation of the Bible.

Noah's Flood has frequently been the object of

such study, some of it legitimate and some not. Claims have been made that the remains of Noah's ark have been found on a mountain in Turkey. More recently, Drs. Walter Pitman and William Ryan have shown that the Black Sea flooded very suddenly during the time we are discussing. Their book, *Noah's Flood,* makes a case that the ancient flood legends of the Western world must stem from this event, which would have sent thousands of people running for their lives out of what was then a fertile valley. They have made a very plausible case that this flood took place.

They also make the point that oral tradition must have been much more important before the use of written language, and indeed, the fact that ancient epics were ordinarily recited, being handed from parent to child to be memorized, suggests that stories might well move down the ages essentially intact. The fact that Turkish storytellers today still recite historically accurate tales, such as the *Epic of Gilgamesh,* from the distant past suggests this.

There is also an interesting connection between ancient myths that has been little studied, although a fascinating book, *Parallel Myths,* by J. F. Bierlein, does compare mythologies from all over the world, as does *Hamlet's Mill*.

These two books reveal that myths are not exclusive to cultures, even if they are separated by vast distances. In fact, the myths of the world reflect common human experience.

The peoples of the world carry with them a number of universal stories. Among these are the myth of human creation, the idea of the fall of man, the story of the hero's journey through the underworld or world of the dead, and, most common of them all, the myth of the great flood.

So the Black Sea may well have been one source of the story. But it was not the only one. Rather, a picture appears of a world that was fractured by a catastrophe that caused universal rain and universal flooding.

It is doubtful that the Incas, the Cree, the Algonquins, the Mojave-Apache, the Choctaw, and many other North American Indian tribes had ever heard the biblical story of the Flood. And yet they all have myths of mysteriously rising waters and a hero who saved people in one way or another. In two of the stories, the bird sent out to find land—the raven—is the same as the one in the story of Noah. While some of these myths were collected in the nineteenth century and may thus have been influenced by Bible stories, the most archetypal of the American flood myths clearly dates from before contact with Europe.

The Aztec flood myth follows the pattern of the other myths of the world: People became so wicked that the gods inundated the whole world with rain. One couple was spared. The man was even called *Nena,* not so different from *Noah.* He was instructed to make a boat, which he did. Unlike Noah, who was favored by Yahweh, Nena annoyed the Aztec gods, who were far stricter than the Jewish deity. Nena was turned into a dog, and the gods started the world with a clean slate.

The Inca myth, like the story of Noah and so many of the others, also relates the cause of the flood to a reaction on the part of heavenly powers to a period of extreme human wickedness, a time of ceaseless wars and barbarism. Of course, myths generally blame the victims for natural events that they could not possibly control. But it is also true that the near-universality of the inner structure of the flood story, with its hero who rejects a wicked world and his subsequent rescue of the animal kingdom and a few human survivors, may well be a memory of a world social order of some kind that was drowned during the catastrophe we have been discussing.

One of the most common themes among the flood myths is that there was a long period of rain preceding the flood. Not a few days or weeks, but months and months of rain. In other words, peoples across the whole planet seem to remember a flood that involved, not a mysteriously rising lake, which is how the Black Sea inundation would have been perceived, but a massive, overwhelming rainstorm.

The way that these and so many of our most ancient myths seem to tell parallel stories strongly suggests that the Greek, Aztec, and Hindu concepts of there having been successive ages of man, the end of each of which left the planet virtually scoured of its previous inhabitants, must reflect some sort of truth.

The sudden methane drop that took place around eight thousand years ago was accompanied by a return to cooling, and it was just as this took place that the past superstorm we have been speculating about probably happened. This storm did not trigger another ice age, but if it had happened in the fall or winter, it might have.

But what made the methane drop? Could the storm itself have caused this to happen? Methane gets into the atmosphere from a number of sources: human activity, volcanoes, rotting vegetation, animal digestion, and chemical reactions like the one that took place in the Mediterranean that we discussed in the last chapter.

Unlike carbon dioxide, methane dissipates if its source is removed. The methane from the landslide under the Mediterranean may have caused enough heat retention to trigger greenhouse warming, but it doesn't explain why the levels stayed high. There is no evidence of the kind of extensive volcanic activity that would have been needed. And it would have taken massive increases in the population of methane-producing animals—creatures that eat mostly grass—for that source to make a difference.

So some unidentified source kept the methane levels high. It had to be steady, and it had to be large enough to sustain what a few thousand years before had become a much-larger-than-average level of atmospheric methane.

There isn't anything known that could have done that. But the methane is there—until, suddenly, about eight thousand years ago, it drops very quickly, back to a more normal background level.

The earliest agricultural sites also appeared around this time. The origins of agriculture have been traced to a number of different highland areas, most notably around Lake Titicaca in the Andes; the central highlands of Thailand, where rice began to be domesticated in the area of Spirit Cave; Ethiopia, where millet was first domesticated; and other regions such as the area of north-central Europe, where agriculturists displaced from the valley that became the Black Sea apparently settled.

It could well be that man did not go extinct because of the invention of agriculture or the transporting of it from flooded areas to highlands by refugees. In any case, judging from the way the ancient stories describe what happened, survival must have been a major achievement.

The early stories describe not only a flood, but some sort of disaster involving the sun as well. The sun is described in many of the North American Indian stories as "rolling in the sky" or "bouncing" across the sky. Days and days of darkness followed, as well as torrents of rain, rising water. Agonizing cold decimated whole populations.

Since then, we have enjoyed thousands of years of warmth, and our species has grown and prospered until it has reached an unprecedented level of wealth and happiness.

However, it is possible that we have reached a moment of crisis similar to the one that caused the last episode of extreme weather. It could be that the same situation that occurred then is happening now.

20

Paris in the Dark

Not since the Germans burst through the Ardennes in 1940 had refugees been seen crowding the roads of France. At first, they had been Germans and Poles and Czechs, then Danes and Belgians, and a pitiful smattering of Norwegians and Swedes. There were no Finns. There were no Latvians, no Estonians, no Russians.

They came swarming in, pouring in, hordes of them in trucks, in limousines, in Trabants, on trains, in planes. France was in

*no way prepared to receive what soon
became half a million refugees a day, then
a million, then an uncountable horde
jamming the snow-blown highways, choking the secondary roads, finally marching
through the fields, walking across icy
rivers, buying, begging, robbing.*

*A desperate French government found itself
completely unable to cope. Defying the
European Union, it attempted to close the
country's borders. But the army was not
trained in border management, and the whole
northern part of the country was quickly
becoming impassable due to the blizzard.*

*In one way, this was a relief to the
government, because the flow of refugees
began to drop. All that remained of them
now were long ripples in the snow—the
buried roads jammed by their buried vehicles. Nobody knew how much death was out
there, but it was a lot. It was a lot.*

*In more northern Europe, the scale of the
disaster was completely beyond imagination.
Sweden was buried under seventy inches of
snow, and it was still falling, in squalls as*

fast as three inches an hour. The temperature was far below zero, the wind gusting as high as 180 kilometers an hour. There were survivors, of course, in the larger cities, on farms with their own generators, wherever human beings could continue to find warmth. Nuclear power plants were magnets for humanity all over northern Europe. Defying regulations, people crowded directly into the facilities, seeking shelter inside when the power grid failed and they could no longer transmit their energy.

The storm continued its relentless, heedless way. France, forewarned by what was happening to the north, decided to defend Paris from the storm. France decided that, given the threat, there was no other choice.

Snow removal equipment from all over the country was rushed to the capital. Most of it came from the alpine provinces. Most of the ploughs were followed by long lines of refugees taking advantage of the briefly opened roads.

These refugees did not follow the equipment to Paris, though. They headed

south, and soon the cities along the Côte
d'Azur were as crowded as if it were mid-
August.

As discreetly as it could, the government
moved its major organizational entities
south, all the way to Marseilles, but the
more visible political institutions remained
in Paris.

And Paris went to war. It was an act of
lunatic courage, a defiance of something so
terrible that it could hardly even be understood, let alone beaten. Nevertheless, the city
set about saving itself.

The storm, which could be perceived as a
single, organized entity only from satellite,
seemed on the ground to consist of an endless series of blizzards, one worse than the
last. These cells would race across a
thousand miles of territory, then break up
like thunderstorms as they either blew themselves apart or collided with geographical
features.

When their cloud tops reached high
enough, a brutal cold-air circulation with
ultrahigh winds would begin, and the land

below would be devastated. When such cells struck cities, as had happened in Edinburgh and Saint Petersburg, the cities were all but destroyed.

Snow was falling over all of France, but for the coastal provinces. Even the Atlantic Maritimes were freezing, because the warm currents that had moderated their climate, making places like Biarritz mild the year round, had disappeared into the new circulation. Now these areas were subject to the full force of the North Atlantic, which had been waiting at bay for thousands of years.

Paris struggled, Paris fought. Aboveground electrical connections and power stations were manned by dozens of workers armed with gasoline-driven heaters. Flying squads of technicians stood ready to reach any point of breach in the power system. One thing was extremely clear: If the power failed, then the war against the storm was lost.

Why was Paris fighting? Because it saw itself as the living symbol of Western civilization, the wellspring of the greatest thought, the repository of the greatest art, the ancient center. It identified its own survival with the survival of the civilization to which it had, in such great measure, given birth.

So even as the long darkness of the storm descended, the lights of Paris continued to glow. Of course, many thousands of Parisians thought the whole business was insane and joined the refugees going south. They were not fools, though, and headed, not for Provence, but for the Midi, where things were sure to be a little cheaper, if also a bit cooler.

A supercell was observed moving toward Paris dropping snow at the rate of three inches an hour, carrying winds in excess of 140 kilometers an hour. This storm struck at a time when there were already forty inches of snow on the ground. Smaller streets had already been abandoned to the storm, and now the emergency task force began to limit its clearing activities to the main thoroughfares.

Paris was cut off from the north, but there were still supply lines open along the

main routes southward. However, the breakdown of order was so profound that
foodstuffs were not being brought north. The
city began to see shortages, and those who
had remained to defy the storm began to
suffer want. Still, though, the city had
power, and as long as that was true, it was
felt that the city would live.

The wind screamed through the girders of
the Eiffel Tower, around the dome of Sacré
Coeur. Immense snowdrifts swarmed down
the Champs Élysées, along the Avenue Foch
and the Rue de Rivoli. The Tuileries Gardens
were buried, the Avenue Charles de Gaulle
became impassable.

Paris, which is situated on a wide, flat
expanse marked by a few stands of low hills,
was wracked by winds that rose higher and
higher. Roofs began to fail all over the city,
their tiles and other materials blowing into
the air like so many leaves.

Aluminum and glass was torn off the skyscrapers of Montparnasse, and office
furniture cascaded into the storm. A mass of
smashed cars, desks, window frames,

awnings, roofs, and all manner of debris began rolling through the streets, choking intersections and forming the basis for mounds of snow forty and fifty feet high.

The métro still ran on underground sections, but the RER had to be shut down. The inner city was now completely isolated, as if it were under siege.

Still, though, Paris persisted in its details. It did not die, not completely. The unexpected continued to define Parisian life, as it had since it had amused Roman overlords with bearbaiting in the Arena of Lutèce. Incredibly, a number of cinemas managed to stay open. Why? How? Nobody could say. But the fact was that the Grand Rex was showing Dien Bien Phu, *a dour 1993 epic about the collapse of French Indochina. Why they had chosen this particular film was not clear, but it apparently made sense to somebody. Studio 28, where Luis Buñuel's* Age of Gold *had been premiered to general outrage in 1930, mounted a round-the-clock Jerry Lewis festival, free to all comers for the duration of the storm.*

Despite the general lack of food supplies, many restaurants struggled on, if only because the proprietors were now unable to leave the city and there was therefore nothing else to do.

Then Restaurant Jules Verne, a quarter of the way up the Eiffel Tower, had a tragedy. A wall of windows was blown in during dinner, and patrons and staff were forced to evacuate, but not without losses. An unknown number of people remained trapped in the rubble and were presumed dead.

Normal rescue services ceased to operate. Bodies all over the city ceased to be counted, then to be moved, then even to be found.

Still the storm increased. There were no further communications from anywhere in northern Europe, except for weak transmissions from what were presumed to be underground military facilities in Russia.

The political institutions of the government were trapped in the city. France began to appeal for help.

Spain, Portugal, Italy, Turkey, and Greece were undamaged, but they could not get relief across the Alps and the Pyrenees. The United Kingdom was suffering even more terribly than France. The British Isles had become untenable from the north downward, pressing millions of refugees ever farther south. But areas normally warmed by the Gulf Stream suffered the same fate as the Atlantic coast of France. The current was gone, leaving the full force of the Arctic free to spread its killing frost.

Great Britain had suffered the destruction of perhaps fifty percent of its population, maybe more. All services had failed. All organization was at an end. The survivors, trapped and freezing, huddled along the south coast, choking desperate towns and cities, starving.

There were no reports from Ireland, from Iceland, from Greenland.

Still, though, the lights of Paris shone in the storm. At Lucas Carton, Alain Senderens devised a cold sweet made of snow and tart citrus; the piano at Le Zephyr was in almost constant use; new designer drugs that made cold seem warm began to be sold in various places.

A cat fur coat cost fifty thousand Euros, an ounce and a half of gold, or a pound of beef. A pair of snowshoes made from tin and old car tires could be had for a thousand Euros.

But snowshoes weren't what mattered if you needed to go out. You had to protect your head and face. You could not risk even brief exposure to the cold, which dropped at night to -70 degrees Fahrenheit.

Satellite views showed no surcease of the storm. But nevertheless, things were changing. The waters of the North Atlantic were getting colder, and the conditions that had spawned the superstorm thus were coming to an end.

Still, though, the winds howled; still the streets of the city became more and more choked with snow.

Power had dropped so far that heating could be continued only in essential environments. In the museums, paintings were taken from the walls and moved into basements. The Orangerie, the Louvre, the Musée d'Orsay—all strove to protect their treasures. The wind shattered windows and swept through bare halls, crusting statuary with ice, making the Venus de Milo glitter like some alien life-form in the darting flashlights of her guards.

But still the lights shone on—flickering now and again, it was true, but not for long, never for long. All heat except electrical heat was now turned off. Oil and gas supplies had failed.

Fires began to be added to the problems of the freezing city, as makeshift heating systems failed and overtaxed electrical wiring burst into flames.

Nothing could be done to control them, and burning buildings simply had to be abandoned. Le Bal du Moulin-Rouge mounted a show called "Les Demoiselles de la Neige," and a few customers actually showed up to watch three elderly Montmartre prostitutes and two transvestites clomp around opening and closing tissue-padded raincoats.

And then the lights failed. It happened in the gloom of the luncheon hour, first a flicker, then darkness. Everybody waited. In the Presidential Palace, in the métro, in hospitals and apartments, in restaurants and shops, in the cinemas and theaters, they waited.

The lights did not return.

21

Mechanism of Destiny

~~~~~~~~~~~~~~~~~~~~~~~~~~~~~~~~~~~

THE SUPERSTORM WOULD BE THE MOST VISIBLE part of the climate shift that moves us from our current temperate period to another ice age. But it is not the beginning of the process. Many things would have to happen before such an event could be triggered, and they would have to happen in exactly the right sequence. This is why the storm—if it happens at all, and we think that it does happen—is so rare.

The sequence of events is this: There has to be greenhouse warming, and it has to reach such an extreme that the Arctic itself begins to melt. The Arctic Ocean has to be flooded with enough freshwater from that melt to cause it to get warm enough so that the temperature differential with tropical waters equalizes sufficiently to cause the current to weaken.

When the current ceases to penetrate into Arctic waters, their temperature will drop. This will cause the tropical airflow to stop and will result in cold air, which has been held in the high Arctic, to plunge southward, colliding with the warm air mass that has moved north. The situation will be exacerbated by the extreme cold of the stratosphere, which will greatly intensify the violence of the storms that result.

If these conditions are met, vast storms will be inevitable. It is possible that they would form into a superstorm. The evidence from eight thousand years ago is that global warming reached an extreme similar to the one that is occurring now. There was a sudden flood of freshwater into the ocean. Already, according to oceanographers in Australia, huge volumes of the oceans have become less salty (*New Scientist,* 31 July 1999, p. 22), and the Arctic is losing an average 26,000 square miles of ice a year (*New Scientist,* 7 August 1999, p. 5).

The Arctic is melting now and it would appear that it will only be a few more years before extensive summer melt takes place and the conditions favorable to a superstorm are present.

When the storm is over, probably within a month to six weeks, the Northern Hemisphere will have sustained enormous damage. In the aftermath, a substantial part of its northern half would be covered with snow, much of it packed to ice, all of it frozen very hard. Depending on the season in which the storm took place, this ice would either form the foundation of another long period of glaciation, or it would melt, resulting in floods of biblical proportions caused by the runoff.

How far are we from this? It seems so massive and so sudden that it's almost impossible to conceive. But we could be just years away. We have seen that the weather has a violent past, so let's take a careful look at what's happening now. And if we can change things, let's find out how.

First, we need to survey the present situation. There are a number of things happening that are possible preludes to the superstorm. The first of these involves the Antarctic. It does not play a direct role in the superstorm, but if vast amounts of freshwater from melting ice pour into the Antarctic Ocean, there will be a subsequent rapid rise in summer water temperatures, making it even harder for the weakened pull from the far north to enable ocean currents to continue circulating as they do now.

At the same time that the ozone layer has been thinning above the Antarctic, dramatic changes in the ice mass are being observed. As early as 1988, huge icebergs began to calve off the Larsen Ice Shelf on the western side of the Antarctic continent, near the tip of South America. By 1998, half of the shelf had broken up and the other half was in danger of melting.

In 1994, a massive section of the Larsen Ice Shelf broke off. This twenty-two-by-forty-eight-mile ice floe was the largest in many years, and prompted Argentinean glaciologist Rodolfo Del Valle to comment, "We predicted that the ice shelf would crack in ten years, but it has happened in barely two months."

The process has continued beyond Del Valle's worst fears. By February of 1998, a massive 160-square-mile piece detached from the Larsen-B Ice Shelf. Dr. Ted Scambos of the National Snow and Ice Data Center in Boulder, Colorado, speculated that the loss of this section could have destablized the surrounding shelf enough to cause its breakup. Scambos said, "Things that had been stable for several centuries are no longer stable." In March of 1998, the Larsen-B Ice Shelf retreated past its historical minimum when another huge mass of ice fell into the sea. Scambos concluded, "This could be the beginning of the end."

The British Antarctic Survey has recorded an extraordinary retreat of Antarctic shelf ice over the past five years. In January of 1995, the Larsen-A shelf had completely disintegrated, with the loss of nearly a thousand square miles of floating ice. Every square mile of ice that melts reduces the salinity of the surrounding oceans, salinity that is crucial to the all-important circulation of the world's ocean currents. Reduction of salinity at the extremes of the currents' flow in Arctic and Antarctic waters is particularly critical.

The reason that this is all happening could not be more straightforward: both poles are getting warmer, and fast. Since 1940, the Antarctic's average annual temperature has increased by about 6 degrees Fahrenheit. The Arctic increase is even greater: 8 degrees.

If some of the thicker Antarctic ice shelves, such as the Ross and the Filchner-Ronne, were to collapse, they would release even more massive quantities of freshwater into Antarctic seas.

If the ice shelves at the edges of the continent melt and break up, then the flow of ice off the sheets will increase. Both actual experience in Antarctica and global warming predictions suggest that this will eventually happen, but almost no experts consider it an immediate threat. Whether that is an accurate assumption or not remains to be seen, inasumch as almost no experts in 1985 were predicting loss of ice on anything like the scale that is taking place now.

In the June 23, 1998, issue of *Nature* a paper appeared by Dr. Michael Oppenheimer that discussed the condition of the West Antarctic Ice Sheet. Dr. Oppenheimer concluded that there was a high degree of likelihood that the sheet would eventually melt, with the result that coastal areas of the world would be flooded. These coastal areas, incidentally, presently contain two dozen of the world's greatest cities and more than a billion souls. However, the paper was reassuring on one point: It was not thought likely that this would happen for another five hundred years. But Dr. Oppenheimer's assumptions did not reflect the rapid changes in ice conditions that are being observed now.

In July of 1998, a disturbing report appeared about one of the glaciers that is a key to the stability of the West Antarctic Ice Sheet. Pine Island Glacier, which flows into the Amundsen Sea, has been retreating for the past four years, according to a report in the July 28, 1998, issue of *Science* by Dr. E. J. Rignot of NASA's Jet Propulsion Laboratory. The hinge line of this glacier—the point at which the ice ceases to be attached to the continent and starts floating—retreated by three-quarters of a mile between 1992 and 1996. If this continues, the glacier will eventually begin breaking up and will flow rapidly into the sea. This is worrisome, because scientists have identified this glacier as the key to the stability of the whole West Antarctic Ice Sheet. If it fails, a vast amount of ice could flow out behind it. Because glaciers are notoriously variable creatures, there is at present no way to predict the degree of danger, but there is little evidence that the glacier will survive for anything like five hundred years and abundant indication that it could start breaking up at almost any time.

The two most important things that preserve our current climate are: the continued stability of the North Atlantic Current, and the temperature of the upper atmosphere. The greater the difference between upper and lower atmosphere temperatures, the more violent our weather. And, as we have seen, upper atmosphere temperatures are dropping rapidly right now, because greenhouse gases are trapping more and more heat close to the ground.

Conditions that will result in a climate shift are building all over the world.

Let's look now at the Arctic ice pack, where in 1997 and 1998 scientists began to observe unprecedented thinning of the ice. Buildings began to topple in Siberia as permafrost melted and shallow foundations disintegrated. In Alaska, melting permafrost began killing millions of trees by flooding their roots.

In September of 1998, the National Oceanic and Atmospheric Administration announced that the previous month had been the hottest August on record worldwide—the eighth in a series of "hottest months." August, NOAA said, "continued the unprecedented string of record-breaking temperatures." It appeared that worldwide temperatures in 1998 were about 1.3 degrees Fahrenheit above normal. Temperatures in Paris touched 100 degrees in August, and New Delhi topped the world's cities at 126 degrees. Overall, there was a far greater temperature increase than had been predicted by all but the most radical global warming models as recently as 1995. The year 1998 ended as the hottest one ever recorded, and in 1999 it became clear that temperatures were rising far faster than had been anticipated just a few years before.

At first, this makes it seem as if we are entering a runaway global warming scenario. Under such a scenario, the earth becomes unable to radiate enough heat, and the atmosphere locks into rapid, unstoppable warming. Temperatures in a few years reach a point where there would be a general collapse of the environment as we know it, followed by heat so great that the planet would cease to be able to sustain human life, or—in the end—any higher life-forms at all. Before this happens, however, it would appear that the climate returns to balance. Prior to the last sudden cooling, Greenland ice caves reveal that a sudden increase in arctic temperature of over 50 degrees took place, probably in only one or two seasons.

The way that the oceans circulate, distributing heat across the planet, is what determines our weather. When the great oceanic currents change, climate changes with them. And because currents change suddenly, so does climate. In 1997, Stefan Ramstorf of the Potsdam Institutes for Climate Impacts Research reported that the Gulf Stream has already been weakened by increased flows of freshwater into the North Atlantic. "There is a threshold in the North Atlantic ocean circulation beyond which the circulation may abruptly collapse." He went on to say that this could happen in the 22nd century, "but it could be much sooner" (*New Scientist,* 14 November 1998, p. 15).

We base our global warming models on the amount of carbon dioxide mankind is releasing into the air. Human activity—running combustion engines, creating power and heat out of coal and oil, and so on—produces this gas, which serves to reduce the ability of the air to release heat it absorbs from the sun.

Human activity currently releases about as much CO_2 in a year as a small volcano might, but unlike a volcano, we never stop erupting. None of our models foresee any unplanned or unexplained reduction in CO_2 emissions, and they all show the amount of the gas in the atmosphere increasing rapidly, as it has been throughout the 20th century.

But it started at a very low level. In fact, for the past 3 million years CO_2 levels in our atmosphere have been incredibly low. Compared to what is usually the case on earth, the carbon dioxide content of our atmosphere is still so low that it has almost no insulating effect at all. In fact, in the whole geologic history of the earth, levels have been this low only once before. That was 300 million years ago, long before the age of the dinosaurs, when underlying conditions were close to what they are now. At that time, a massive ice sheet stretched over what is now called South Africa and world climate was harsh and cold.

In the past, whenever there has been a return to ice conditions, it has been preceded by a rise in greenhouse gases, followed by a sudden drop. The drop might reflect the sudden shift in climate we have been discussing.

Right now, we are somewhere along the path to this change. Stratospheric temperatures are dropping. Surface temperatures are rising, especially in the Arctic.

In early 1999, reports came back from scientists who had spent a year aboard the Canadian ice-breaker *Des Groseilliers* that the North Polar cap was melting with unexpected speed. Indeed, the destruction of the northern ice appears to be even worse than what is happening in the Antarctic.

The project aboard the *Des Groseilliers* was called SHEBA, short for Surface Heat Budget of the Arctic, and involved 170 scientists measuring the condition of the Arctic ice. As we have seen, the more this ice melts, the lower the salt content of the surrounding water and the more susceptible it is to warming and cooling over short periods.

In the 1970s, ice thickness in the Arctic averaged three meters. What happened in 1997 was described by SHEBA chief scientist Donald K. Perovich: "The first problem we had was trying to find a floe that was thick enough. The thickest ice was 1.5 to 2 meters." In other words, in just twenty years, the ice has lost nearly half of its thickness.

The Arctic Ocean, which is shallow and thus apt to change rapidly if it is flooded with freshwater, was found to be warmer and less salty than twenty-two years earlier. This means that a large amount of ice had melted before the summer that SHEBA started. As of March of 1999 some scientists are beginning to conclude that much of the Arctic ice cap will be open ocean, at least in the summers, within a very few years. It appears that it will lose as much as seventy percent of its ice mass within another twenty-five years.

In addition, in March of 1999, it was announced in the journal *Science* that the Greenland ice sheet was diminishing. Like the Antarctic ice sheets, the Greenland ice sheet is on land, and its melting will therefore cause a rise in sea levels. Worse, the same sort of glacial surge that is a possibility in the Antarctic could occur here, resulting in extraordinary flooding worldwide, and massive amounts of freshwater entering the oceans.

On previously unsurveyed areas of Greenland's eastern ice sheet, the ice had thinned by nearly eight inches a year for the past five years. Closer to the coast, thinning was taking place at the rate of thirty-six inches a year. It appears that Greenland's glaciers are flowing into the sea much faster than expected, raising the possibility that they will surge suddenly into the ocean.

Dr. Gerard Bond of Columbia University's Lamont-Doherty Earth Observatory has said that the increase in freshwater entering the ocean from more icebergs would result in the types of ocean current problems that we have been discussing. Dr. George Alley said that the possibility existed that "putting more freshwater into the Atlantic would cause things to change in a hurry." According to the *New York Times* of March 5, 1999, the suggestion was that the effect would be "as if a light switch were being thrown: A little pressure may not cause the system to change, but when the pressure reaches a certain point, it flips suddenly."

This process has been amply demonstrated by the examples from the past we have been discussing, which offer much evidence of sudden change.

Dr. Alley said that scientists "have no clue" about how close the switch is to flipping.

So the following conditions exist for sudden climate shift and the superstorm:

1. Surface air is trapping more and more heat because of the greenhouse effect. This is causing the upper atmosphere to get colder. The more extreme the difference in temperature, the greater the violence of the weather.
2. The Arctic Ocean is getting less salty and warmer as polar ice melts and iceberg flow increases.
3. The Antarctic ice pack is melting as well, flooding the South Atlantic with freshwater.
4. Ocean currents are weakening.

What does all of this add up to? When will the superstorm take place?

Frankly, we don't know. When the climate switch flips, it's going to be very sudden.

If a warm snap melts enough Arctic ice fast enough, the stage will be set for the collapse of the North Atlantic Current. When this takes place, the type of climatic chaos that we have been speculating about in this book will occur. As things now stand, this situation is inevitable. It is going to happen, and nobody knows when.

Will it be strong enough to trigger another ice age or just cause a great catastrophe? Again, nobody knows. Can we do anything about it? Fortunately, yes.

22

Hope and the
Human Future

~~~~~~~~~~~~~~~~~~~~~~~~~~~~

EVEN AS WE HAVE COME TO THE CLIFF'S EDGE
of the superstorm, we have reached the most in-
candescent moment of knowledge in the history of
the species. Staring down into the abyss of destruction, we break through into the heights of understanding.

On dozens of different fronts, human knowledge is expanding at a pace that could not have
been imagined even in the past few years. To look
back to the seventies or the eighties is to peer into
a strange, ancient world, slower and smaller, with
horizons so narrow that they now appear almost
laughably constricting. In 1985, you went to a
library to do research. You could not fly on a plane
without significant expense. A capable computer

filled a room. There was no Internet. In fact, our world was far less than it is now, and not only in these ways. When we looked to the future, even the future of science, we could see a clearly defined borderland. Not only that, the environmental situation had a hopeless feel to it. There was a sense of gathering doom, although not one of great danger as is true now.

Most of the predictions that were in place in 1985 showed massive population overgrowth, the expansion of pollution, and the rapid decline of species. And we were entirely responsible. Rather than being just one factor in a great and complex extinction event, we were seen as the sole cause. As a result, all of our environmental planning was impeded by feelings of guilt. We were somehow *at fault.*

Nevertheless, mankind did respond to this environmental crisis.

And this is what we accomplished:

In 1975, solar energy cost seventy dollars per watt. By 1997, in constant dollars, this had dropped to four dollars per watt. World military expenditures dropped from a peak of a trillion dollars in 1988 to 700 billion in 1996. Wind power cost twenty-six hundred dollars per kilowatt in 1981 and was down to eight hundred per kilowatt in 1998.

In every area, from the production of pollutants to the growth of population, mankind has been an amazingly successful environmentalist. Defying all but the most optimistic predictions, the population of the earth has grown far more slowly than anticipated in 1985.

In short, just as the environment is challenging our very existence, we are responding with a massive, worldwide, and heartfelt effort to survive, and this has happened despite the fact that the environmental policy of the most powerful country in the world is hobbled by a false debate about the need for it.

The same "critical mass" switch that flips the environment suddenly from one state to another operates in human society. Just as the environment is reaching a negative threshold, human civilization is reaching a positive one that may prove to be of equal or even greater power.

Part of the change is social. It is almost as impossible to remember how the geopolitical world actually looked just fifteen years ago as it is to recall the state of science and technology then.

The U.S.S.R. was an unassailable political reality. There were few who seriously questioned its long-term viability. As far as the politicians of the era were concerned, some sort of centrally planned Soviet reality would continue to govern Eastern Europe for the foreseeable future.

Within a very few years, Russia was a federation, Eastern Europe was free, and Soviet Communism had collapsed. Contrary to every expectation, a threshold had been reached, a switch flipped. A gradually bending social situation had suddenly snapped.

Since then, Eastern Europe has embarked on a program of modernization and self-help that is almost unprecedented in history. We hear, for the most part, about the troubles in the Balkans. But the reality of Eastern Europe is that one of the most environmentally troubled areas on earth is beginning the process of a massive cleanup that will, in the end, result in a complete restructuring of an aged and deeply poisonous industrial infrastructure.

Still, Eastern Europe is a tiny island in an ocean of irresponsibility. Asia is almost totally lacking in environmental consciousness, beginning with China, which is engaged in a process of self-poisoning so profoundly destructive that much of the Chinese countryside could become spoiled for human life within a chillingly short time, climate change notwithstanding.

Latin America over the past fifteen years has experienced a massive migration both into its capital cities and its hinterlands. In their effort to feed and house their exploding populations, nations such as Brazil have gone to war with their tropical forests, attempting to establish human colonies in places that cannot survive human encroachment intact.

They have set the stage for a catastrophe almost as great as that which appears to await China.

So, despite all the success of the recent past, we need to do better. We need a breakthrough.

Like all breakthroughs, it will not be expected. It will come from an area now considered part of the fringe. It will consist of either new knowledge or rejected knowledge that has been rescued by a visionary.

A scientific breakthrough that might cause such change would be something like viable nuclear fusion, the development of an efficient and environmentally friendly way of storing electricity, or the discovery of a means of removing pollutants such as CO_2 from the atmosphere.

All of these areas offer considered reason for hope. Progress is even being attempted in the esoteric area of extracting energy from the vacuum of space, promising a limitless supply of energy so abundant that it could even propel us into outer space in numbers significant enough to matter to the earthly environment.

Such a breakthrough would lead to mankind exploding onto the frontier of space in meaningful numbers, reducing the pressure on the earthly environment. Pressure would also be reduced by the development of a fusion reactor, or better, a fuel cell robust enough to power a home. Both of these technologies are tantalizingly within reach. Further removed, there is a possibility that a strange experiment conducted by Nikola Tesla shortly after 1900 could result in a new source of energy. The experiment involved extracting electrical energy from the ionosphere. There is some evidence that the technology was suppressed by oil interests, but it would appear that its return, if it is workable, is all but inevitable in a world as needful of clean energy as this one.

Science has never offered mankind greater promise than it does right now, nor have we ever been in a better condition to take advantage of it. Even as they gamble with environmental disaster, our economies are evolving into fabulously efficient engines of health, wealth, and happiness. Limitless promise beckons across the frowning gray horizon. On the one hand, we have China's average personal income increasing by four hundred percent in real wealth in just twenty years. And on the other, we face the problem of what would happen if each Chinese citizen began to consume as much oil in a year as each American does. China would, in that case, require 80 million barrels of oil a day—more than the total current world production of 67 million barrels a day.

So China can't grow . . . or so it would seem. But China is packed with inventive, hopeful, and brilliant human beings, just like the rest of the planet, and if their growth is thwarted in one direction, they will find room in another. It may be that the Chinese will never consume that much oil. In fact, it will be. Something else is going to intervene. And yet, if history is any measure, the average Chinese citizen will be far better off in another twenty years than now. Somehow, he will have a car. Somehow, he will have food, television, ice, education, linen, a computer, and all else for which he strives.

And somehow, his world will be polluting less than it does now. But always there is that nagging question, Will it be enough? Will we manage to step back from the edge of the cliff?

We have seen how the continual shifting of the planet's ecosystem back and forth between ice ages and warming periods has served as a challenge to us, demanding again and again that we change or die, sink or swim, evolve or end up on nature's ash heap.

We are at that point again, but this time, instead of simply reacting blindly, we are able to anticipate the coming catastrophe and, if not avoid it, then certainly do what we can to survive it in the best order we can manage.

People see all around us the evidence of rapid climate change. All over the world, violent weather has made its indelible mark. Despite the best efforts of the global warming skeptics, the open and obvious fact of environmental decline is a far more effective propaganda tool than any lie or obfuscation cooked up by environmental skeptics, and citizens displaced by savage and abnormal storms are a more effective voice for change than any Washington lobbyist pleading to preserve the *status quo* can ever hope to be.

The result of this situation is a hidden environmental revolution taking place all around us right now. For example, between 1990 and 1997, the use of fossil fuels worldwide increased just one percent while the population rose twelve percent. During this time, solar cell sales increased at the rate of fifteen percent per year. Then in 1997, some sort of a threshold was reached. A switch was tripped as people all across the Third World discovered the freedom that solar cells could bring them. They could have electricity and fuel. They could have the life they dreamed of. And suddenly worldwide sale of solar cells jumped forty percent in a single year.

Wind power has been growing at the amazing rate of twenty percent per year for the past ten years. In fact, wind is the key growth energy industry in the world, not gas or oil. Not only that, the potential for wind is fantastic. Just three states—Texas and the Dakotas—according to the Department of Energy, possess enough useable wind to supply the whole electrical demand of the United States.

The fantastic promise of zero fossil fuel use seems within the reach of existing technology.

Similar changes are taking place beneath the surface in the transportation industry. Numerous countries and cities are seeking means to free themselves from the automobile as the foundation of transportation. But individuals are voting with their feet—literally. Over the past ten years, world car sales increased 60 percent to 37 million vehicles a year. But world bike sales soared a stunning 424 percent, to 106 million units in 1997.

Populations are no longer growing uncontrollably. Fourteen percent of the human population now lives in areas where there is population stability. Forty other countries, containing more than half of the rest of the population, now have fertility rates of under three children per woman, including China. This means that world population stability is a believable possibility, and there is every indication that the rate of increase will continue to decline, defying all past predictions.

Mankind wants to survive. We want to prosper. And that is true not just for the elaborately comfortable individual in the First World country; it's true for those in poorer areas as well. People can see that preserving their immediate environment is important to their continued health and increased prosperity.

We appear to be at the edge, at the beginning of the twenty-first century, of an entirely new kind of revolution, one that will have the same sort of effect on our lives that the industrial revolution of the early nineteenth century had on the society we are in the process of leaving behind. The environmental revolution of the next millennium will proceed hand-in-hand with scientific advances that will actually make environmentally sensitive economies more prosperous and more pleasurable to inhabit than the old economies, which are indifferent to their effect on nature. Consumers across the planet, especially in the rising Third World, will come to see environmentally supportive modern methods of generating power, providing transportation, and increasing their food production as more wealth-producing than the old ways.

This vision isn't at all idealistic. Who wants to spend hours every day gathering wood and dealing with cooking over a fire or pay for fuel when solar cells can accomplish the same thing with less cost and effort?

A huge number of human beings are racing against time to fit themselves into a violently stressed environment before it becomes incapable of handling them.

In 1985, the average person was deeply ignorant about the environment. Now, with the rapid spread of world literacy and education, this is no longer true. The people of Honduras, for example, who endured devastating floods during Hurricane Mitch in 1998, understand, generally, why this storm caused them such extreme suffering. They know that farming techniques that are too intensive, that leave new earth exposed, and that are indifferent to the dangers of runoff lead to devastating floods. They have seen almost their entire country washed away by floods of almost inconceivably destructive fury.

Will they go back to the farming techniques that so amplified the hurricane's damage? Given the burgeoning demand for cultivation advice among Honduras's *campesinos,* that seems doubtful. It is likely that the next storm to strike Honduras will be met by a farming community much more prepared to handle its runoff effectively.

In other words, the collapsing environment is doing what it has always done to us, throughout our history. It is challenging us to greater achievements and teaching us unforgettable lessons in survival.

Seen from this perspective, our unstable environment is the essential wellspring of human evolution and change, and the more serious our environmental problems become, the more creatively we are responding to the challenge.

The future that gazes back at us from across the great environmental divide of our era is a strange and wonderful one, a place well worth going. This wonderful, complex, fascinating, and endlessly varied world civilization that we are creating is, for the most part, a very enjoyable place in which to be human. For many of us, even now, the world in which we live seems not only fun but also important. Our lives delight us, but they do more. As never before in human history, we are tasting the pleasures of the spirit. Reading is a growth industry worldwide. In every functional country, the libraries and bookstores are filled with jostling, striving people, each one intent on feeding his spirit in this ancient way. At the same time, film and television programming are becoming more sophisticated than they have ever been before, and in every country in the world, the choices are expanding almost limitlessly.

New technologies are exploding all around us, to the point that no futurist, no matter how savvy, can meaningfully or reliably predict what our world will be like in just ten years, let alone fifty or a hundred. The two of us are in the same situation. We are well aware of the fact that future reality is continually outstripping present imagination.

Staggering advances in miniaturization and computing sciences, for example, promise that we will soon be able to embed ourselves with information, adding areas of expertise, even memories, even dreams, to our own minds with the same ease that we now add software to our computers. As we come more and more to understand the mechanics of thought, we are also reaching toward the seat of the soul itself, on the point of discovering the all-important difference between intelligence and consciousness and, at the same time, striving to be within just a few tantalizing years of building an intelligent machine.

When we do this, the melding of man and machine will become so intricate that where the human personality ends and the machine personality begins will be a diminishing shadow line.

In response to the problems we face right now, we are going to find ways to generate and apply ever greater intelligence. Just as the development of more and more powerful engines defined the twentieth century, the development of ever more powerful minds will define the twenty-first.

There will be machines more intelligent than we are, machines that can evaluate our needs and understand our problems better than we ourselves can. At that point, many of the mysteries that currently defeat us are going to be solved. For example, the inability of science to predict future climate will undoubtedly come to an end, and with it the argument about how to plan usefully for a healthy human future.

If we survive, the great quest of the twenty-first century is going to be for machine intelligence. We will find out if an intelligent machine can ever become conscious; we will find out the meaning of the soul. And in so doing, we will at last come face-to-face with the true value of human creatures . . . just when we are finally rising beyond what nature made, into a new kind of humanity, self-conceived and designed around an unimaginably powerful combination of man and machine.

In other words, we are going to attempt, on the eve of the superstorm, every single thing we can in order to survive. Society may achieve the unachievable. The tools we don't have, like a working model of the climate, people are going to invent.

Why are we so certain of this? Because it is what humankind has always done in the past when confronted by the dangers of its cyclical climate. We have been struggling against nature since the beginning of our being, and this struggle is about to go into overdrive yet again, because we are at once awash in promise and facing disaster, conceivably even extinction.

So we're going to fight, just as we fought a hundred thousand years ago when the ice came back, and eight thousand years ago when, in its death agony, the ice age almost drowned us all.

The last time nature betrayed us, we invented agriculture. If there was an old human order, it passed away, leaving behind only enigmatic remains that even now cannot be duplicated, reminding us that, along with the gains that the pressure of change brings us, there are also losses.

One thing we obviously lost was the knowledge that led our forefathers to leave us with the riddle of Hamlet's mill in the first place. Whoever devised the myth of the mad miller and his flawed mill, and all the related stories, sent a coded message right to our doorstep. Is it a message from the mirror of a too closely observing eye, or did a real civilization actually perish when the miller last went mad?

What matters is the essence of the message: something happened then that was violent and dangerous and destructive, and there is every evidence that it happens frequently in the great cycle that grips us and is shortly going to happen again.

Just as our freezing ancestors learned to make clothing and organize themselves into social groups in order to survive, we must learn new techniques in order to guide ourselves past the perils of the next few years.

We need more reliable models of climate, and perhaps we need intelligent machines to help us make them. The problem of the weather has to be solved. Prediction must be made so nearly perfect that policy can be constructed on a foundation of knowledge being stymied by the sort of debate we must engage in now, which has the effect of forcing us to gamble with nature, an approach that is obviously extremely dangerous.

There is, however, another frontier we can cross in order to help ourselves, and this one is much more earthbound and practical. This is the frontier of awareness. There has been a sufficient amount of climatic upheaval and the changes, especially in the Arctic, are rapid enough now to suggest to us in the strongest terms that social and political policy needs to be changed. But policy is being constrained, maybe fatally, by what has become a shortsighted debate.

So, we need to get personal. We need, each of us, to understand what we can do, and do it on our own.

If each person in the developed world did no more than seal his house against air leakage, insulate his water heater with a cover, and run heating and air-conditioning thermostats at 68 degrees or below in the winter and 74 degrees or above in the summer, carbon dioxide emissions would be reduced significantly enough to measurably slow down global warming. Slowly at first, and then with increasing speed as these simple habits spread, weather instability would begin to decrease. It wouldn't be enough, of course, to solve the problem, but it *would* buy us some time.

In addition, we can make an effort to encourage car manufacturers to build more energy-efficient cars like the Toyota Prius. Using a combination of a gasoline engine and an electric motor, this vehicle has virtually zero emissions and gets up to seventy miles per gallon. It does not require recharging and has about double the range of the average car. It's built on a Corolla body and has more than adequate speed, pickup, and as much room as a normal Corolla. In short, this and the other cars in this new class are no-compromise vehicles that are not only environmentally friendly, but also more convenient than normal cars.

At the same time, Ford has just introduced the largest passenger vehicle in history, a massive nine-passenger machine that dwarfs even the Chevy Suburban. Built on truck standards, it is a mega-polluter and gas-guzzler. It joins the ranks of dozens of other sport-utility vehicles, all built to meet the lower standards applied to trucks, and all spitting out far more pollution than normal cars.

To work on behalf of the future, we need to vote with our feet in the showrooms and stop buying these vehicles. We need to insist that our legislators change current law and place SUVs under the same emission standards that govern passenger cars.

If we start with personal change, then an entirely different future comes into focus on the horizon. In *this* future, our magnificent civilization survives, its prosperity spreading across the world in a sustainable manner. By 2050, mankind will have broken through into a dimension as yet unguessed, will probably have spread into outer

space, and will be generally healthy, happy, educated, and full of hope for the future.

We are at the edge, treading on the wire. Will we
cross or will we fall? Our ancestors fell. Let's heed
their warning. This time, let's make it across.

23

Why Doesn't Somebody *Do* Something?

~~~~~~~~~~~~~~~~~~~~~~~~~~~~~~~~~~

OUR REFUSAL TO DO ANYTHING MEANINGFUL about our role in sudden climate change is a gamble, and what we are gambling with is the largest, richest, and most valuable thing that mankind has ever produced: our civilization. Too many things are building up right now that push against the switch, threatening to throw it. If that switch flips again, we are likely to go the way of our forefathers, leaving behind us only myths and fading memories.

The history of our climate over the past 3 million years confirms that we are sitting on a powder keg—even without human interference. Our activities are speeding up everything, which probably means that when change does take place, it's going to be extremely sudden. It's not simply a matter of flipping the switch this time. This time, it will be thrown with a truly violent hand. If we do nothing, it's going to happen. When? Given the way things are changing, especially in the Arctic, the danger is already with us.

Whether or not a new ice age will be the result depends largely on when the superstorm develops and how long it lasts.

As we have seen from the condition of the fossil remains the last superstorm left behind, it happened in summer. The presence of a budding apple tree, quick frozen and left for thousands of years entombed in the permafrost, confirms this.

The fact that the storm happened in summer is why it caused, not an ice age, but a mass of floods so extensive that they are remembered in practically every culture on earth. At some point, there was a melt of the snow the storm had laid down.

We are, in fact, taking a great chance with a tremendous danger. So why don't scientists warn of the danger? Why don't they raise the alarm? The two of us are amateurs. We don't think that we're wrong, but serious scientific work has to be done before we can be sure of our ideas.

To inform mankind, with authority, of how serious the danger is, there must be more work done on things like the effect of global warming on the flow of ocean currents. It is scientists who must make the predictions on which changes in social policy depend.

The problem is that scientists can't predict future weather, not with enough certainty to enact the kind of dramatic policy changes that are needed if we're right. So we're in trouble. Scientists might think that sudden climate shift is a danger. But unless they can provide us with a model that proves it, nothing will be done on a governmental level.

For decades, it has been hoped that a reliable model of large-scale climate change would be developed. But model after model has failed to provide real-time results that are convincingly predictive.

One of the problems is that the amount of detailed climatic data available drops to almost nothing further back than a hundred years ago. Prior to that, records for the most part are too sketchy to allow for any kind of modeling. So most climate models start with the past few years, when there has been enough data to work with. Nowadays, information from surface stations, wind profilers, aircraft, Doppler radar, and satellites can be combined to build a clearer picture of the weather. But for even regional climate change predictions (as opposed to short-term forecasts) to be made correctly, the Department of Energy says that the ability of the models to handle data must improve tenfold.

Meanwhile, the official world is going to remain unable to act. No irrefutable scientific warnings are going to be sounded. And yet, it is clear that the clock is ticking. It's very clear.

Because science can't produce certainty, propagandists and politicians have effectively stopped any kind of official-level reform that might save our world.

In 1988, Dr. James Hansen of NASA's Goddard Institute of Space Studies announced that a process of global warming was taking place primarily because approximately one ton of carbon dioxide was being emitted into the atmosphere every year for every human being on earth.

Immediately, oil and chemical companies, some oil-producing countries, religious fundamentalists, and American political extremists responded. Ever since, these forces have been spreading a consistent message: nothing is proven, so we should wait and see. In 1998, the U.S. Congress even went so far as to attempt to prevent government officials from speaking publicly about global warming, in order to derail American participation in the Kyoto Conference, the most recent worldwide attempt to address the issue.

The oil industry maintains a propaganda machine called the Global Climate Coalition that spends big money to spread the "wait and see" message. The National Coal Association does the same. The National Petroleum Institute retains a public relations firm to help defeat taxes on fossil fuels. The NPI alone—just one part of the fifty-four member Global Climate Coalition—spends nearly as much as all the major environmental funds put together.

OPEC, the consortium of oil producing nations, has joined with big oil companies such as Arco, Exxon, Sun, Shell, and Unocal to spread the word that fossil fuel emissions should not be controlled.

One informational film created for this purpose is called *The Greening of Planet Earth,* a video produced by the highly activist coal company Western Fuels. This film claims that increasing carbon dioxide in the atmosphere will actually *improve* our lives by enabling us to turn current desert areas into grasslands. The theory is that more atmospheric carbon dioxide will make plants more robust. The flaw in this idea is that deserts are not caused by a lack of plant growth, but by geographic conditions such as mountains that deflect moist currents of air and by soil conditions that prevent growth. So increasing carbon dioxide is not going to make the deserts bloom. Instead, as any midwesterner who endured the floods of 1996 or Chinese who suffered through the Yangtze floods of 1998 will tell you, increasing carbon dioxide is going to cause deluges in places where rain already falls normally. Worse, carbon dioxide is the primary greenhouse gas at the moment, and its buildup is the center of our current problem. Despite the fact that levels are low in terms of geologic history, they are still causing enough heat increase to threaten the kind of polar melt that is potentially so dangerous.

The fuel industry, backed by foundations with deep pockets, maintains a cadre of individuals with university degrees who skillfully drain the issue of all sense of crisis. Congressional conservative Republicans offer staunchly partisan support, dismissing all talk of this universally important issue as "liberal" posturing. Of course, it really has nothing to do with political ideology. The issue should never have been politicized in this way. It is an issue above politics.

However, the propaganda causes people to feel that they must side with ideological liberals if they are to take a position in favor of solving our environmental problems. It is perfectly appropriate for people of any political persuasion to take the stance that these problems must be solved—especially those of us who have children.

The American public has been misled into taking a "wait and see" attitude, one which amounts to a gigantic gamble, a test of the ability of nature to tolerate punishment.

Unfortunately, the situation has not been helped by the inability of our scientific institutions to respond to the problem with the kind of decisiveness and clarity that is demanded. Even as we see massive climate change all around us, science itself is trapped in this devastating but unavoidable debate about its predictive abilities.

Even given the recent attempts to create better models, there are fundamental flaws in the whole modeling process that render science even more impotent.

The so-called General Circulation Model, one of

the primary tools of climate research, does not adequately explain the way that carbon dioxide is exchanged between the oceans and the air, with the result that it cannot be used to definitively prove that global warming is under way. The skeptics point out that global average temperatures have changed only slightly in the past fifty years, during which period carbon dioxide emissions have increased many times over. Science cannot adequately answer them.

The problem goes deeper even than the lack of good models and the tools to use them. Even the way science is organized is detrimental to aggressive and clear prediction. Science is divided into thousands of very narrow specialties, for which reason findings tend to be reported in isolation. This means that, while individual pieces of information may be persuasive, they are rarely integrated into a big picture. The result of this is that enormous questions such as whether the planet is moving toward another sudden change in climate don't get the kind of clear and unambiguous answers we need in order to act.

Without an irrefutable model around which to organize this data, there appears to be plenty of room for debate. Unfortunately, this is probably a dangerous illusion, given that the last time a superstorm blew up, animals seem to have been killed so fast that they were still chewing their food when they froze solid. We are playing games with a climatic disturbance whose impact is worse than that of a full-scale nuclear exchange and, in some ways, just as quick.

Since science hasn't got the ability to be definite and there is a large, effective, and well-financed group on the other side of the equation, there is virtually no chance that we are going to act decisively as a society in this matter.

So the key becomes to recognize when the stage has been set for the superstorm and to act radically at that point. The hope would be that it wouldn't be too late.

The stage becomes set when winter temperatures rise to an unseasonable level above the Arctic Circle while the salinity of the Arctic Ocean is dropping. At that point the North Atlantic Current will begin to change course. Such an event will trigger many smaller storms before the superstorm itself begins. In fact, the presence of dozens of smaller storms might even obscure the larger pattern. We don't know exactly how a superstorm looks, after all.

But once it gets started, the superstorm will not stop until the energy driving it is dissipated, possibly not until the North Atlantic Current returns to its normal course.

Any effort at weather control during the formative stages of the storm might be futile, but if it was possible to derail it, what would be necessary would be to radically reduce atmospheric heating in the temperate zones of the Northern Hemisphere. This would cause the Arctic Ocean to become cold enough to reestablish the North Atlantic Current earlier in the storm's progress, thus removing the primary engine.

The difficulty would be to realize in time that we were in a superstorm situation. Right now, weather reporting tends to be local and regional, only occasionally transcontinental, never planetary.

But even if we had warning that the superstorm was beginning to gather, we would not have the means in place to act. Worse, the forces that fight any kind of environmental reform, as well-funded and powerful as they are, would interject the same note of caution that they now use to keep us passive and undecided.

So it is possible that we might not respond to the storm even as it formed. We might not do the things that would make a short-term change: stop driving cars, use only minimal electricity, damp all possible fires, reduce the amount of airline activity. We might instead increase our use of heating oil and gas as temperatures dropped, then attempt to migrate en masse out from under the storm's path as the wind began to rise, emitting even more CO_2 as we did so.

We have never witnessed anything as violent as the superstorm, and people can be expected to panic when it is realized that it represents a highly unusual situation, an environmental breakdown of the highest importance and greatest danger.

A large number of major cities, from Vladivostok to Toronto and including most cities of north and central Europe and the northern tier of American states, would quickly get into serious trouble. Sustained winds in excess of a hundred miles an hour would trap people indoors or in their vehicles. In the Arctic and subarctic, ultracold microbursts might freeze whatever they struck in a matter of minutes. There would be a growing atmosphere of panic as satellite communications were interrupted by the heavy cloud cover and the infrastructure began to falter. Power failures would begin to take place early in the progress of the storm, further impeding people's ability to communicate and, finally, to survive the extraordinary cold.

As the storm continued, power stations relying on the delivery of coal would shut down, as deliveries would stop. Oil and gas transmission would be interrupted as exposed pipeline sections burst or pumps on oil lines burned out due to the increasing viscosity of the cold oil.

At some point, failure of the power grids of most affected countries would occur. Desperate migrations would ensue, as people realized that survival inside the storm's path was, at best, problematic. But roads would be blocked, snow removal having failed along with gasoline deliveries. Food supplies would fail for tens and, finally, hundreds of millions of people.

In London and Paris and Moscow, in New York and Toronto, the lights of civilization would begin to flicker and dim. As the storm's intensity peaked, the number of structures collapsing would reach astronomical levels, as wind pressure and snow weight combined to create stresses never imagined.

It is possible that the death rate among people trapped inside the main body of the storm would approach one hundred percent. Left living would be only the few and the lucky—if they could be called lucky. Those surviving until the sun came out, probably sometime in mid-March if the storm started in early February, would face a vast, glaring desert of ice, a slick, cracked, treacherous plain swept by high winds.

The major centers of civilization would, essentially, be ruined. The educated minority of humankind would have been cut by as much as twenty percent in a matter of weeks. The highly organized system of wealth distribution that has been created over so many painstaking generations would be no more.

In the United States, the southern border of the storm would lie somewhere south of Kansas City. If any part of the federal leadership remained alive, they would probably be headquartered someplace like Atlanta.

The U.S. might have lost a third to half of its population. Canada, Russia, Finland, Sweden, Norway, Iceland, and Scotland would essentially be dead. The British Isles would be a devastated remnant. Of European nations, only Spain, Portugal, and Italy would be likely to remain intact. If the storm did not result in permanent freezing, the resultant spring melt would flood every river system in the Northern Hemisphere.

Not until you reached Mexico, in the Americas, and North Africa would you begin to find states that were still undamaged. Southern nations like Brazil, Argentina, Mexico, and South Africa would become the new focus of the human future. Hong Kong and Singapore would be the financial capitals. Australia and New Zealand, like Japan, would have received damage, but not from the central process of the storm. They would be likely to recover normally, at least in the short term.

The time of year that the storm took place would determine whether or not its aftermath consisted of massive floods or the beginning of a new ice age.

Floods of truly epochal proportions are described in the literature of myth from Pacific Coast Indians to ancient Sumerians. Many tribes of the American Great Plains, such as the Mandaw, Choctaw and Knisteneaux, describe a huge flood. In the Knisteneaux account it is so tremendous that it actually turns the prairie into an ocean, rising to the point that virtually the entire population dies. That the bones of mammoths, rhinoceroses, and other animals are found congested at high points on the plains where they died suggests that the Knisteneaux and other such stories are true, and very ancient. Animals that would have been frozen farther north would, in this area, die of exposure and, eventually, drowning.

If the ice did not remain frozen, the flood that would follow a modern superstorm would be similarly devastating. Almost unimaginable destruction would occur south of the already desolated storm line. Rivers like the Mississippi would briefly become sheets of water hundreds of miles wide, drowning everything in their path, rendering escape impossible for large populations that would be scarcely able to imagine what was happening to them. Fuelless, out of communication, their entire social infrastructure broken down, the people of this ruined region would not be able to organize themselves against the new danger of the flood any more than they had been able to cope with the storm.

In the event that the ice did not melt, survival in the areas the storm had not touched would at first be more certain. The southern tier of American states would probably regain their organization in a reasonable length of time. Southern California would be able to draw on its traditional sources of food in the Imperial Valley, and most of the Old South would be able to feed itself, although the shortages of corn and wheat products from the devastated Midwest would mean that seasonal shortages would become commonplace and bread would be in short supply.

Famine would spread quickly through most of the world that is now dependent on North American grains. The corn supply to states like Mexico would be interrupted. The Caribbean nations would suddenly find themselves forced to rely on local foodstuffs, and the lack of fuel would put the ability to remove meaningful populations from the islands in jeopardy. Island populations throughout the Northern Hemisphere would be in serious trouble.

Elsewhere, starvation would lead to migration, and large numbers of Mexican citizens would begin appearing in the southwestern United States, with a consequent rise in violence as panic-stricken U.S. citizens fought to preserve what was left of their community and their food supplies.

As the storm would have unfolded along a fairly rigid line of demarcation, the difference between areas affected and those untouched would be great. Areas near the Rockies, the Alps, and the Himalayas would be choked in snow—essentially killed. Russia, Canada, and much of the northern U.S. and China would have come to an end as viable places of human habitation. First World countries like Singapore that now exist in the shadow of their larger neighbors would come to dominate geopolitics, simply by default.

As the years passed and the ice mass grew steadily larger, however, human activity would be constantly and increasingly impeded by ever worsening weather conditions. In the end, the climate of places like Louisiana and Spain would be like the climate of southern Siberia is now. Mankind would be adrift in a sea of memories and recriminations, struggling desperately for survival in an increasingly hostile environment.

Unfortunately, ice ages last much longer than warming periods, and human destiny would now change profoundly. If the scientific knowledge of the present could be preserved and expanded, we would undoubtedly embark upon a plan of understanding and ultimate control of the weather.

Eventually, if the weather was ever understood well enough, a further plan to improve the situation might be launched. This would involve the greatest engineering effort in human history, nothing less than the separation of two continents now linked by a land bridge.

A portion of Central America many miles wide might be excavated by a human species armed with a clear understanding of how the weather actually works. Mankind would then reestablish the ancient equatorial current and attempt to return the weather of the planet to the balance that it had enjoyed through most of its history.

The new current would flow much as currents did in archaic times, around the equator, bringing even heating and regular seasons to the planet once again. The ice would melt, and a much diminished but wiser human population would gradually reclaim the world lost beneath it, the myths and the memories, the glory that was our present age.

Epilogue

New York

It was so white, the land. And the sky was so very, very blue. The air had a purity to it unlike anything Bob Martin had ever known. At –63 degrees, care had to be taken to breathe, but he was well-equipped, feeling something like the astronauts must have felt on the moon.

As leader of the three-man Emergency Damage Assessment team assigned to locate survivors in Manhattan Quadrant 4-A-2 (Fortieth to Forty-fifth Streets, between Fifth and Seventh Avenues), his job was to find heat sources lurking in the ruined landscape, which were possible signs of life. The plan was to use infrared scopes of a type Bob was familiar with from his days in the military.

He had taken this job because it was important. He'd been a Ranger, after all, and done long-range recce (reconnaissance) during the Gulf War. The last time he'd used one of these things, it had been to pinpoint Iraqi sentries, and he knew how to get the maximum response out of the notoriously touchy equipment.

He'd also come here for another reason. He'd lived through the storm in comfort. Martie and the kids were safe in Austin, where he'd been able to buy a condo ten days before prices went through the roof. The pleasant, three-bedroom unit they'd paid $235,000 for would sell today for a million or more.

So he was also here because he and his family had survived. Not untouched, of course. Nobody in the world was untouched.

Martie kept a paper copy of the last E-mail from her brother in England, would always keep it close to her. He'd died at his desk, no doubt of it, assembling satellite data for the Met, the British Meteorological Office.

Any survivor with rescue skills was morally obligated to use them, as he saw it, especially if he was as lucky as Bob.

As he and his two teammates moved slowly down Fortieth Street on their snowshoes, they peered into the fifth- and sixth-story windows they were passing. The glass was gone, all of it. Inside, the rooms were choked with snow.

New York's power grid had failed in the first few hours of the storm. The city had tried to pull together, but then the central steam system had also failed. Three days into the storm, oil deliveries had ceased. More than 2 million New Yorkers had escaped down the system of turnpikes and interstate highways that wound south through Washington and Richmond, where the worst of the weather ended. The roads had been kept open eleven days into the

storm by incredibly brave, incredibly determined work crews. They had saved millions
of lives, these people, often at the cost of
their own.

Nevertheless, it was thought that as many
as a million people might be entombed on
the island of Manhattan alone. Maybe all of
them were dead. Probably so.

But what if some weren't? What if even
one was still alive?

There were six emergency teams like this one
in Manhattan, five more working in the rest of
the city. So far, they were the only units that
had been organized. Thousands were needed,
all over the country. Cities like Philadelphia,
Saint Louis, Kansas City, Salt Lake—all of
them could conceivably contain survivors
whose time was running out right now.

In the absence of any central authority,
the governors of the surviving states had
formed a temporary governing council. The
U.S. Fifth Army in San Antonio had been
redesignated as Continental Army
Command. It had deployed most of its
available resources along the Mexican border

because of fear that looters would invade the part of the U.S. that was still intact.

Nevertheless, it was anticipated that a substantial rescue force could be assembled within a month. National Guard units would join volunteers to comb the whole middle tier of states.

Farther north, there was thought to be no point.

Teammate Mike Guare called out from nearby, "Heat." Slowly, Bob turned, his snowshoes making him ungainly. He dared not slip, because the snow in this street was fifty feet deep, and there was no way to tell what kind of debris might be hiding in it ready to slice a hand open.

With antibiotics now reserved only for life-threatening infections, that was a serious problem. To get cured nowadays, you had to be dying.

Hurrying as best he could, Bob moved closer to the young pilot. "In there," he said.

They climbed through the window, find-ing themselves in what seemed to be a clothing business of some sort. There were rows and rows of men's suits hanging from frozen racks. The walls still had clipboards attached to nails, and a computer system sat on a desk in a cubicle. At the back of the room, snow as light as talcum powder made eddies on the gray-and-black linoleum of the floor.

All three men had their instruments out now, and all three of them were watching heat coming from a canvas hamper full of rags. Could somebody be living in there? A child, maybe? Dr. William Hanks, a Marine M.D. who'd treated frostbite in Greenland and Antarctica, stepped closer to the hamper. "Can you hear me?" he called, his voice muffled by his ski mask.

They had orders to identify, triage, stabilize if possible, and move on. But if they were going to do even that, they had to at least see the person. Gingerly, Bob pulled at the hard-frozen top layer of rags.

It came away easily, like a lid—and suddenly the room was filled with shrill noises and racing movement. Mike yelled as Bill fell back against him. And then they all saw it: *a very thin, very outraged squirrel was sitting on the shoulder of one of the suits.*

"Must be from Bryant Park," Mike said as the creature regarded them with its dark, wary eyes. Mike would know; he'd grown up in Manhattan.

They went back out, and Bob radioed the finding to their base camp, which had been established on the hard snow that packed Central Park's Sheep Meadow. The teams had spread out from the two choppers that had landed there. Right now Mary Travis and the other helicopter pilot would be deploying the tents, setting up the field kitchen, and getting things ready for the return of the teams.

"Two o'clock," Bill Hanks said. At four, they would start back up Sixth Avenue, past the shells of the skyscrapers, toward warmth and rest. Right now, however, they had other work to do. They were responsible for some large buildings, one of which was the main branch of the New York Public Library. It was thought that refugees might have assembled in large spaces. The city's emergency plan had never been enacted, but some people might have known about the stores of food that had been stashed in such places during the second week of the storm.

When they went back out, Bill stopped for a moment and looked downtown toward the soaring tower of the Empire State Building. Bob followed his eyes. Against the heartbreakingly pure blue of that sky, the building was incredibly beautiful. Its windows might be dark and broken, its famous radio tower gone, but it soared above the ravages of the storm in a way that made Bob feel proud and determined.

As a climatologist, he knew the grim mathematics of the situation. If this mess melted, this city was doomed to be destroyed over the next six months by a flood unlike anything previously known. Snow from upstate would soon inundate the city. At some point, most of these skyscrapers were going to fall, their foundations stripped right to the bedrock of the island by the unimaginable volume of water that was coming.

If the snow did not melt, then next year

*would bring more of it, and the year after yet
more. It would not stop, it would never stop,
not in the future, not even in the far future.*

*He stared up at the Empire State Building.
It had been part of the background of his
world, something you showed the kids when
you brought them to New York. That was all.
But now it seemed sacred, sacred to the memory of man.*

*They went on, moving toward the library,
not really expecting to find much. They did-
n't go fast, because you couldn't. They each
had four Powerbars, their only food until they
returned to camp. Reaching Fifth Avenue,
they turned north. Using a grid of the city on
the readout, they kept to the center of the
streets. Huge snowdrifts leaned against many
buildings and covered others, creating the
appearance of a hilly area scattered with
improbable clusters of tall buildings. It would
obviously be dangerous to walk those drifts,
lest you fall through into the rubble of the
structures they had crushed.*

*The library was a disappointment. Except
for some parts of the roof and the upper ten*

feet of the portico, it was swamped with snow.

They got no heat readings from it. Inside, there was nothing but millions of dead books. Bob tried not to think about them. He got on the horn. "No joy, Forty-second and Fifth. Library's a ruin."

He took a deep breath. And stopped. He took another. "Guys," he said, "do I smell smoke?"

The other two began inhaling. Bill pulled off his face mask. His eyes closed, he sniffed. Then his dark eyes met Bob's. There was no need to say anything.

The three of them started toward the library. As soon as they did, they all saw the same thing: a wisp of white curling up against the blue. They began to hurry, lifting their snowshoes high.

As they came closer to the massive front of the library, they found a well-worn path that went down under the tops of the columns and disappeared into the dark interior. The area was blackened by smoke. There were steps carved in the hard-packed snow.

They were all excited. Bob could feel his heart racing. He started into the space.

"Hold off," Bill said.

Bob's military training came back to him. They had weapons, and now was the time to be ready with them. He unsnapped his parka and made the butt of his pistol accessible. "Okay," he said. "Back me up, Bill," he continued. "Mike, hold the position until I call you in."

With Mike waiting behind them, he and Bill moved into the dark. It was smoky under the enormous stone eaves and dark enough that they had to pull back their snow goggles and use their flashlights.

"Listen," Mike said.

Voices—lots of them. Shrill, excited.

"Jesus," Bob murmured.

They continued on. Ahead of them was a broken skylight, beyond it the interior of the library itself. Bob went to the opening. Flickering light could be seen from within— the light of a fire.

When Bob looked down, he saw what was without question the most amazing sight he had beheld during the storm or at any time in his life. Sitting on the wide marble floor below were about twenty children. With them was a young woman. They were all filthy. They were cooking around a fire made of books.

"Hey," he yelled.

Instantly, their chatter stopped. Faces turned. Faces looked up. A young voice cried out, "Daddy!" And all twenty children came running toward the ladder that connected their lair to the outside world.

"Children," the young woman cried, "stay back!"

They stopped instantly. This girl had taught them to obey her with the precision of a drill team.

"Line up," she said. "Go to the reading room."

"It's cold in the reading room, Sister," a voice whined.

"Do it now!"

Then Bob had reached the floor. The woman came hesitantly forward, her eyes

huge in the gloom, her face narrow, her left
hand lost in a mass of rags.

"Can I help you?" she asked.

"We can help you," Bob said gently.

She burst into tears.

Slowly, seeing her being held in Bob's
embrace, the children came forward. There
were nineteen of them, the entire fourth
grade class of Saints Peter and Paul
Elementary School, Far Rockaway. They'd
come to Manhattan on a library field trip.
When the snow had gotten bad and the bus
hadn't shown up, she'd called the school
and been told to wait. The driver had said
he was on his way.

The bus never came. The library emptied
out. The telephones failed. The electricity
failed. At first, she'd gone across the street to
a restaurant and holed up there. But the gas
had gone off, and she'd decided to move
back here where they could use books for
heat.

She was Sister Rita O'Connor of the
Sisters of Divine Providence. None of her
kids had been in touch with their families

since the storm started. They were tired; they were homesick; they were scared and cold. She had one case of walking pneumonia, two cases of frostbite, and her own hand, which was frostbitten and, she suspected, gangrenous.

Bill took a look at it. It wasn't gangrene, but it was a mess.

They radioed the camp about the situation. It seemed that New York wasn't dead after all. The other teams were all reporting pockets like this, pockets here and there.

Quantico decided that the refugees who triaged as "in need of medical attention but recoverable" should be moved to the Sheep Meadow. A C-130 would pass over with a pallet before six. It was en route from a similar drop in Washington. The pallet would contain tents and food supplies for fifty people for three days. Tomorrow, a field dressing station, Arctic issue, would be dropped in.

So Sister Rita and the Saints Peter and Paul fourth grade started off on their last hike. She cried as she moved along beside Bob. She had not thought help would come.

She didn't understand what had happened. In her mind, the blizzard was isolated in New York. When she heard how much of the world was involved, she was quiet for a long time.

The kids were extremely well disciplined. They helped each other. They helped Sister Rita.

When they got back to camp, darkness was falling. Manhattan was without light, so the stars were coming out. The moon was a sliver of purest silver, following the sun into the orange west.

Mary had a tent up, and the air smelled of hot rations. Wieners, Bob thought. The kids were excited, eager to get at the food.

When they entered the one tent that had been erected, they had a surprise. There were at least thirty people already there, jamming the space. They were so ripe that Bob practically gagged, but he managed to suppress it.

They were people from every sort of situation and background—apartment dwellers, people who'd been caught in the city unawares like Sister Rita and her class, people who had sheltered in the subways, in buildings, wherever they happened to be able to find enough warmth and supplies to stay alive.

Most of them were starving. All of them had respiratory complaints as well as the frostbite. A few would soon be dead.

But there was about them something that Bob had read about but never actually seen, the grace and determination of the survivor. Even the children revealed a gentle, persistent strength and a sense of compassion for those around them.

Bob saw what it was that enables human beings to survive in terrible situations, what it was that had made Sister Rita somehow manage to keep nineteen children alive on a journey through hell, what had enabled all these others to endure. The huddled backs, the sallow faces, the quick eyes—he could see the strength; he could see the rock-hard determination in every one of them.

There was a sudden roar and then the sound of a plane's engines dwindling quickly away. Everyone who could bolted for the doors of the tent.

Night had brought brutal, unimaginable cold. It was probably –90 degrees by now. But it was still and absolutely clear. There were no pollutants in this air. The stars almost seemed to sing in the sky—they were so many; they were so bright.

And then Mike cried out, cried out and pointed. All around Central Park there are buildings. It is surrounded by cliffs of apartments. And everywhere, in one window and then another, candles were being lit. People had heard the planes, they had seen the tents, the lights down there and they were responding.

The candles were so many that it looked as if the stars had come down. There wasn't a sound, but the candles could not have spoken louder: We're here, they said, more of us than you thought possible, many more.

We're still here.

Afterword 2004

by Whitley Strieber and Art Bell

FOR THE FIRST TIME, AS THIS EDITION OF *The Coming Global Superstorm* is published, there exists the possibility that there could be extremely violent weather associated with one of the key changes identified in the book.

In late 2003, the Woods Hole Oceanographic Institution published a study indicating that the polar oceans were losing their salinity while tropical waters were becoming more salty. At the same time, dramatic changes in ocean currents crucial to the stability of the world's climate were being recorded, and unusual and violent weather was unfolding worldwide. Woods Hole scientists have detected a change in the Gulf Stream circulation of the North Atlantic "of remarkable amplitude" that is "the largest and most dramatic oceanic change ever measured in the era of modern instruments."

The failure of the Gulf Stream will be the greatest catastrophe in recorded history. It will not spare a single soul on earth. Everybody will be affected. There will be no "safe areas." Nobody can be certain of the extent of the damage this disaster will do, but it will be extensive and it will last until the current starts again. The horrifying truth is that we could remain in the new climate regime for thousands of years, during which the ability of the earth to support life will in effect drop off a cliff.

Sudden climate changes are part of Earth's climate pattern. They have happened many times over her history, and on a regular cycle in the past three million years. The rising of the Central American land bridge 2.8 million years ago changed the fundamental structure of the ocean's currents, which, combined with variations in solar energy output and slight changes in Earth's orbit, has led to a much less stable climate since then.

There have been numerous ice ages lasting approximately a hundred thousand years each, interrupted by interglacials lasting from ten to fifteen thousand years. All of human history has unfolded during the last third of the most recent interglacial.

The next ice age will begin when the ocean's currents cease to pump warm water into polar regions and snow melt fails to take place across a series of summers. This will increase the surface reflectivity of the planet and reduce temperatures enough to cause glaciers to begin to form over wide areas.

Prior to that time, there will be a dramatic warming spike, followed by a violent change in weather. This warming spike almost unfolded during the summer of 2003, as temperatures in Europe soared to unprecedented highs, tens of thousands died, and even Arctic temperatures rose far above normal.

Man cannot alter the overall cycling back and forth between ice ages and interglacials, but proper planning can minimize the impact of climate change on civilization and possibly delay and reduce the violence of climate change. This can be done without significant cost to business or government.

However, world governments have been unable to agree on a unified planetary policy to delay climate change, despite the fact that a simple plan that does not involve any significant financial commitment is available. If world leaders got together to support the Canada Challenge, vast CO_2 emissions reductions could be achieved, especially in developed countries that can most easily control their emission levels. The challenge is not to industry or government at all, but to the end user of energy, the individual whose everyday decisions affect how much CO_2 will be emitted. Its details are described elsewhere in this book.

If the Challenge were to be accepted by just 20 percent of U.S., Japanese, Canadian, and European inhabitants, world CO_2 emission levels would drop to a point that the human factor would be vastly reduced as a source of global warming. We would gain needed time to prepare for the inevitable.

Unfortunately, there is an almost total lack of leadership worldwide, led by the United States, which has adopted an outrageously dangerous posture of aggressive indifference to climate change at a time when billions of lives are at stake.

Once the Gulf Stream fails, the earth's climate as we know it will also fail. Following the weather upheavals that accompany the failure, a new climate will unfold. It will not be anywhere near as congenial to human life as our current climate. The result will be upheaval on an apocalyptic scale.

No continent—indeed, no city, town, or person—will be untouched. The worst of it is that this could now happen at any time. After it happens, those who ignored the danger but were in a position to do something to help will be counted among the most evil human beings who have ever walked the earth. They will be responsible not only for the death of billions, but for causing modern civilization—the best, freest, and most valuable civilization that mankind has ever created—to pass into history.

ACKNOWLEDGMENTS

Anne Strieber contributed her editorial skills to this book, as well as her insights and skeptical, searching questions that were so helpful in the long process of research and writing. We would also like to thank Werner Reifling of Paperchase Press, whose early faith in the project encouraged us to continue; our agent, Sandra Martin, whose vision took it beyond the limits of our own expectations; and Mitchell Ivers, our editor at Pocket Books, who has supported us throughout the process of bringing our message to the public by his own inspiring belief in its importance.

INDEX

INDEX

INDEX